Lorna Gold works for Trócaire, the Irish Catholic Development Agency. She is a specialist in International Development who has worked across academic and NGO contexts for almost two decades. She writes and speaks extensively on Pope Francis' teaching in *Laudato Si'*. She holds a PhD in economic geography from Glasgow University and currently lectures in applied social studies in Maynooth University. She has published widely, being the editor and author of numerous books, journals, policy reports and academic papers. She is a regular contributor to public debate on climate justice in Ireland and internationally and helped co-found Stop Climate Chaos Ireland in 2006. She is vice-chair of the Global Catholic Climate Movement and a member of the Irish government's advisory group on the national climate dialogue. She is originally from Scotland and is married with two children.

D1428525

CLIMATE
Generation
Awakening to Our Children's Future

LORNA GOLD

VERITAS

Published 2018 by Veritas Publications
7–8 Lower Abbey Street
Dublin 1, Ireland
publications@veritas.ie
www.veritas.ie

ISBN 978 1 84730 841 2

10 9 8 7 6 5 4 3 2

Extract on p. 168 from 'Hieroglyphic Stairway' by Drew Dellinger, taken from *Love Letter to the Milky Way*, Ashland, OR: White Cloud Press, 2011. Used with permission.

A catalogue record for this book is available from the British Library.

Cover designed by Barbara Croatto, Veritas Publications
Printed in the Republic of Ireland by Walsh Colour Print, Co. Kerry

Veritas books are printed on paper made from the wood pulp of managed forests. For every tree felled, at least one tree is planted, thereby renewing natural resources.

'Leaving an inhabitable planet to future generations is, first and foremost, up to us.'
Pope Francis

For my boys

CONTENTS _____

Foreword   11

Acknowledgements   15

Noticing   17
Beginnings   25
A Wake-Up Call   35
A Heavy Heart   47
One Problem, Many Solutions   55
More than the Statistics   71
Heartbreak   81
Sleepwalking   95
Embracing the Earth   109
Change at Home   125
Story-Changers   137
Planetary Movement   151

Epilogue Roar   167

FOREWORD _____

There are many reasons why science has failed to sensitise the person in the street regarding the urgency of tackling climate change. Lack of political leadership, powerful vested interest groups, and communication failings, are among the usual excuses expressed to explain why the most pressing problem of the twenty-first century does not occupy a more prominent position in the consciousness of the average person. But perhaps a less well publicised failing is an inability to bring the issues down to a personal level, a level which impinges on the everyday life experience of people, causing them to question their priorities, and consider disrupting their conventional and sometimes comfortable way of life.

Such a journey of transformation is related in this text by Lorna Gold. Drawing on her life experiences from childhood in the shadow of a Scottish oil refinery to motherhood in an Irish country town, she positions herself as a parent picking up danger signals for the future of her two children. In a highly personal account she builds a poignant picture of someone who has wrestled with her emotions as she embarked on her journey of awakening regarding climate change. The contradictions between mother and activist, the need to see beyond the day-to-day mundane priorities, the growing awareness that time is running out – all are forensically examined in a narrative that is compelling to read. The language is straightforward and accessible and always deeply personal. One cannot but share with the author her growing acceptance of the conflicting emotions swirling around her as she seeks to give her two young sons a wider experience of a natural world increasingly absent from the formative years of most children in the developed world. Yet there is no guilt trip in evidence here. Certainly the injustice of climate change insofar as it impacts on those least able to bear the burden is well explored, and the author's background in the Irish Catholic Development Agency, Trócaire, makes her ideally qualified to comment on this. But there is no moralising from a height in this text. Rather there is a cry for understanding and acceptance that we cannot continue on the present path as self-centred, consumerism-obsessed individuals and imagine we are protecting our children's future. The book opens with the analogy of a lioness protecting her cubs and warning off any external threats. Like the lioness, the author's cry of warning becomes louder as the text proceeds, exploring the political perspectives of Naomi Klein and Bill McKibben and the faith-based environmentalism of Pope Francis. Eventually the cry becomes a roar that seeks to enlist all generations, including the delightfully named 'raging grannies'!

This is an eminently readable book which flows seamlessly across a multitude of themes, ranging from the scientific to the psychological. The plain language grabs the reader throughout and captures the essence of the message very successfully. It was obviously a catharsis for the author personally to come to terms with the conflict between the treadmill of everyday life and the need to radically change the organisation of society to tackle climate change. But maybe it is a catharsis we all have to undergo eventually, and in writing this book Lorna Gold provides insight into the turbulent and transformative journey that lies ahead.

JOHN SWEENEY,
Professor Emeritus, Department of Geography,
Maynooth University

ACKNOWLEDGEMENTS _____

I'm indebted to so many people for their ideas, encouragement and expertise in making this book a reality. A huge thank you to everyone in Trócaire with whom I have worked closely over the last two decades, especially to Justin Kilcullen and Eamonn Meehan; to Professor John Sweeney, Fr Sean McDonagh and many other climate scientists who helped me grasp the problems we face and also the solutions; to the many climate activists young and old, who have inspired me along the way; to Michael O'Connell of 'Take Stock' for mentoring me and encouraging me to follow my dream; to my many friends who read drafts and helped me get to the finish line; to all my family near and far, especially Mum, for being such an important part of the story; last but certainly not least, to my husband Kevin, for all his love and support.

NOTICING _____

Many years ago I went on safari in northern Tanzania. I hired a local guide and took a trip to the Ngorongoro conservation area, a world heritage site which is home to the colossal Ngorongoro crater, an impenetrable fortress some six hundred metres deep. From the outside, it looks stark and hostile; yet its exterior hides one of the most beautiful places on earth. As we reached the summit of the winding, precarious road to the crater's edge, we were greeted by Maasai warriors dressed in their traditional red checked garb. Beyond them, the vista was a vast African plain, brimming with all kinds of life. A vision so perfect, its existence seemed almost implausible. Elephants, giraffes, baboons, wildebeests, hyenas; every animal, large and small, was cradled in this Noah's Ark. Towering, fat baobab trees decorated the lush savanna. Standing there on the edge of the crater was like walking straight into the opening scene from *The Lion King*.

As we drove down the steep track, bumping our way to the plains below, we spotted a pride of lions. What luck to come across these majestic creatures! The lions were sprawled about, lazily digesting the morning's prey and enjoying the early morning sunshine. Behind them, a few yards away, stood the hyenas, ready to move in when the lions' share was gone. Behind them, circling high above, were vultures, ready to swoop down once the hyenas had had their fill. These awesome creatures looked so different in their natural environment compared to the glass-fronted enclosure in a zoo where I was accustomed to seeing them. They exuded power and elegance in equal measure. Content among their own, they had a familiar gentleness about them as they relaxed together and played.

As I watched on silently, I spotted a lioness with a newborn cub. I felt deeply privileged to witness something so rare and intimate. Seeing my excitement, my guide wanted to take a closer look. We eased the jeep forward, trying to close in for a better picture. I almost had the perfect shot when we heard a few low growls and heads began to rise from their relaxed slumber. I had already taken enough photos. But my guide probably felt his tip was dependent on the perfect shot, so he continued to edge in, closer, closer, closer. Suddenly, the lioness, who had been gently tending her young cub, sprang to her feet. Tail pointed upward, back arched, she stared straight at us and made a terrifying roar that pierced the still, hazy air. Echoes could be heard right across the crater, as the warning bounced around its walls and returned to us amplified. Instantly, all of the lions were then on their feet, all of them roaring, all of them staring at us; the intruders. We knew we were no longer welcome. Our skidding tyres sent red dust flying.

The roar of that lioness still rings in my ears and reverberates in my whole body. It was a life lesson in respect for the power of nature and our vulnerability to other beings, especially when we are guests in their habitats. It still reminds me never to outstay my welcome and to be conscious of the warning signals around me. Above all, however, that incident speaks to me of something awesome and awe-inspiring that I also share with the lioness: a powerful instinct to protect my young from danger.

I really only understood that instinct, the fierceness and intensity of that love, when I had my own children. It is something so strong, deep, visceral, that it can overtake me. It can manifest itself at the most unexpected moments. One day I was out with my two boys who were riding their bikes. My youngest son was four and still trying to master the art of balancing on two wheels. We arrived at the top of a path winding down through a park beside a river. His older brother saw the challenge and freewheeled down to the bottom of the path at speed. Eager to copy his big brother, my younger son followed suit. However, at a certain point he lost control and came off the path – careering down the bank and headed straight for a gap in a hedge which led to a river. I was a few steps back. When I realised what was happening I yelled something, a strange cry, as I seemed to watch the scene in slow motion. I ran like I never ran before, adrenalin filling my whole body. Those few seconds seemed to last forever. I caught him just before he reached the hedge. I was shaking, terrified, relieved and held him so tight. Oblivious to the danger, he just smiled at me and told me he was fine. That night, I lay awake. I could feel the pain of those few moments of fear, of that protection instinct taking root.

Like the lioness, I will do anything to protect my young from danger. And that is why I decided to write this book. You may well ask what on earth writing this book has to do with protecting my kids. I am writing this book in the hope that perhaps someone will read it – maybe you, maybe someone else – and understand a little more about what is happening to our world, particularly due to climate change and the real and present danger we are placing our own children in.

You certainly don't need to have your own children to understand what this book is about. Most people can grasp the kind of instinct to protect that exists between parent and child. We have all been children. All of us have, or have had, parents and grandparents. The vast majority of us have at some point been on the receiving end of that fierce love. It is the kind of love that instinctively sticks up for a child who is being bullied or drops everything without thinking at the sound of a certain cry for help. It is a love that would go to the ends of the earth if a child were sick, in the hope that somehow they might get better. It is the kind of love that gives the tightest hug and doesn't want to ever let go. It can be fierce.

This book is my personal story of waking up to the danger our children are in. It asks how we can protect them before it is too late. Just like my son on his bike, I fear their future is speeding out of control; and unless our protective instinct kicks in, we may lose what is most precious. We may do so unwittingly, without so much as a thought, often in pursuit of the best for our children. Yet we all have that instinct to protect deep inside us; it's in our DNA. It is essential to the survival of our species. Above all else, understanding climate change now is about coming face to face with our children's future and that protective instinct. It is about safeguarding them against future harm. It is about cradling their dreams. To do that, we need to start joining the dots and see the connections that exist between our climate, our actions and their future.

The actual moment when I decided I just had to write this book, however, didn't come to me in any exotic place. It came to me while I was sitting upstairs on a bus, stuck in traffic on Dublin's north quays, on my way to give a talk about climate change in a church hall. It was an atypically mild, sunny afternoon in December. I remember the precise moment: I was looking out the window at The Croppies Acre, a memorial to those who died in the 1798 Rebellion.

As I was staring into the distance, my mind a thousand miles away, I suddenly noticed the garden was full of daffodils. I love daffodils. I think there is something so joyful about them, something that signals the end of winter and the start of spring. These were *not* the normal tiny green shoots peering up above the cold winter soil. No, these daffodils were fully grown, the crowns of yellow petals welcoming in the sunshine. Wait a minute … rewind there. What did I just see? Yes, daffodils in *December*. As I looked again, I realised what a peculiar sight it was. There were no leaves on the trees, the sun was setting in the early afternoon, yet here were daffodils in full bloom.

I turned around to the young man next to me as if to say 'how weird is that?' But he had his headphones in, so I kept quiet. He would have probably thought I was a bit cracked anyway. I looked around at the passengers on the bus.

Everyone was peering at their phones, each in a little world of their own, so removed from the very peculiar scene outside the window. I wonder if anyone else even noticed the daffodils that day or was it just me? So I sat and thought. And thought. By the time I arrived at my stop, my mind was abuzz.

It seemed to me that those daffodils were trying to tell me something. I felt as if they were crying out and no one could hear them. They were silently screaming two words which none of us really wants to hear: climate change. An extraordinarily mild winter – the mildest on record – had caused them to bloom three months early. Everyone was talking about the weird weather, but nobody was joining the dots.

Thoughts were whirring around in my head: Why is it, I thought, that such an important issue is falling on so many deaf ears? How come we can't seem to grasp what is blatantly happening right under our noses and bring ourselves to take it seriously? How come we are ignoring what may well be the biggest threat for the future of the planet? Given that there are so many good people in this world – mums and dads, grannies and grandpas, aunts and uncles, sisters and brothers – who all care deeply about the children in their lives, is there anything I can do to change this situation? I thought about the lioness and her instinct to protect. And that's where the idea for this little book came into my head.

For me, trying to make sense of what is happening to our world has resulted in two essential sides of my life – being a parent and being an activist – colliding and becoming intermeshed. My activism has become an expression of my parenthood and vice versa. While campaigning, I often think about what my children, and their children, will ask me when they grow up – 'What did you do about the climate?' 'Did you know what we would face?' 'Where were you when we needed you to stand up for us and our future?' These questions are so vivid in my mind, they keep me awake at night. They push me to take action and keep going when things seem hopeless. As a mother, I am constantly thinking about what I am teaching my children about our world – judging how to be truthful about what is happening, but also how to protect and nurture them, in the knowledge that their ability to deal with an increasingly uncertain future is in my hands.

And that is what this book is all about. It is my personal story of how I came to understand that certain things are happening to our world which are putting the prospect of our children having a peaceful and happy future at risk. There are of course many such risks to their future, but one problem stands out: climate change. It stands out because it is so huge. It also stands out because it is fixable, yet we choose to ignore it and hope it will go away.

This dawning took time, but was pierced by some startling light-bulb moments. It took place over a number of years and continues to this very day. It has been like piecing together the bits of a jigsaw puzzle. If I am totally honest, I fought it at first. Such was the magnitude of the story and the need for an urgent response, my first reaction was to turn away and hide behind the sofa. It is far easier in fact not to know. If you are fortunate enough to be living in a solid house on high ground in a rich country, ignorance is bliss where climate change is concerned. So is there a reason to keep looking?

The conviction that I need to do my bit to tackle this issue made me come to a personal decision. Not an easy one but a clear one. I might as well spend my short time here on this planet trying to do my part to help protect my children's future and that of the millions of children in the world. I may fail miserably, and probably fail each day, but I think it is a gamble worth taking. It is the only logical response to the feeling that as a parent and a citizen I need to raise my voice and do something to change things. And so I write with a sense of urgency, a plea to be heard.

BEGINNINGS _____

My interest in what is happening to our planet and to people in the developing world has been a driving force throughout my life. I came to Ireland in 2002 to work on these issues for Trócaire, the official overseas development agency of the Irish Catholic Church, but my interest goes right back to my teenage years. Like a whole generation of aid workers, I came of age to the soundtrack of 'Do They Know It's Christmas?' and 'We are the World'. Together with Bob and Bono and their Live Aid concert, we were going to feed the world. No, better – we were going to save it. I still remember when I was thirteen I mobilised my entire school to collect wellington boots for people in the Philippines after a cyclone! After that episode the headmaster rather embarrassingly dubbed me the 'school conscience'. My thinking on how to help people has moved on somewhat from those days.

When I was around fourteen I did a geography project which really made me think about the environment around me. I still remember the smell of the project book, with its white padded wallpaper cover and spiral binding. Back in those days there was no Mr Google to ask, not even a basic computer to type up my project. Every project was like a piece of detective work. I loved it. This particular project was on my local environment. I lived in Scotland, right next to the Grangemouth oil refinery near Falkirk. Every evening our homes would be illuminated by a fluorescent orange glow from the flares burning off the excess gas arriving from the North Sea oil fields to the refinery and ICI (Imperial Chemical Industries) chemical plant. I wanted to research whether the oil refinery was having any detrimental effects on the environment.

In the course of my project I came across this idea that scientists were discussing but nobody was quite sure about. It was called 'global warming'. I remember drawing the earth, the layers of the atmosphere and explaining the 'greenhouse effect' caused by gases in the air that trap energy from the sun. Based on my research I ended by explaining that fossil fuels, and hence the oil industry, were having a major effect on the environment but that those effects were not very clear yet. Acid rain, however, was a problem.

It was a simple project, but one which had a profound effect on me. I remember drawing a timeline underneath it and in my head doing some calculations around when global warming would really hit; around 2020 I reckoned. But that seemed a long way off. Sting would have saved the Amazon rainforest by then and we'd all be fine.

This concern for the environment kept growing. I had become deeply fascinated by the workings of the UN Commission on Sustainable Development and its report 'Our Common Future' in 1987. The report was the result of a UN commission set up to examine the whole idea of how we could live more ecologically. One of the major conclusions of the report was that if we continued to live the way we were living, we would require up to five planet earths to give everyone a standard of living like ours. I was so taken by this report I went out and bought a special issue of the *Scientific American* magazine dedicated to it and read it from cover to cover. I covered it in notes and fluorescent pink highlighter pen. I was gripped by a sense that humanity was on a collision course with the environment. We were all living way beyond our means, while others were struggling to survive.

I was quite a serious child. In fact, at times I felt the whole world was on my shoulders and I had to carry it. Looking back now, it is evident to me that my deep sense of responsibility was in large part dictated by the events of my early childhood. My first six years were, to all intents and purposes, fairly uneventful. My dad was a stockbroker in Glasgow and went out to work each day. My mum was an accounts assistant in the same firm, but gave up work to look after us when we were young. By the time I was six, there were four of us – my older brother Kenneth, me and my two younger sisters, Geraldine and Anne-Marie. Life was pretty good. We were happy in school, Granny came on Fridays and cooked sausages for us. We got a treat at the market on Saturdays. Dad cooked up a big roast dinner on Sundays and we sat by the electric fire on a shaggy pile rug after our bathtime watching *The Muppets* and *Starsky and Hutch* on our new colour TV.

Everything changed in the course of twelve short months. My grandmother, who had been a huge part of our family life, died suddenly of stomach cancer. Meanwhile, my youngest sister was born with whooping cough and was critically ill in hospital. Six months later my dad was killed in a car crash whilst he returned alone from a family holiday. To top it all, just six months later, on the first anniversary of Granny's death, while my mum was out for the evening to mark her passing, we had a terrible house fire. We escaped with our lives. The house was only salvaged thanks to the quick thinking of my brother and our neighbours, but life would never be the same again.

The trauma of these exceptional circumstances had an impact on me, beyond perhaps what I even now understand. People often ask me 'how did you react? How did you cope?' I reacted and coped as six year olds do. I continued to live in the moment and tried to make sense of things in my own way. Yet when it got dark, for years I would wake constantly with night terrors. Walls would fall in on me, fire would rage around me, dead people would come out of graves. Most of them were very nice dead people, thank God, and I often asked Mum when I could go 'up to heaven' and be with Dad and Granny.

Yet, for the most part, we all normalised the new situation – whatever that new normal was. For me it meant caring for my two sisters who were two and one. It meant looking out for my mum – not physically, but doing everything to make her smile again. God forbid anything I or anyone else did would add to her pain and distress. I think I made a subconscious promise to myself at that time that whatever I did in life, it would make her smile. That childhood promise still has power to move me. It is astonishing how our childhood lives buried, often barely so, just under our skin.

I had to grow up fast. As I say, the reality of the events of that terrible year, which from an adult viewpoint screams of heartbreak and distress, was for me the new normal. I earnestly wanted to make things better and I learnt very quickly to change nappies, to feed my sisters, to bath them, mind them and to play with them. I helped out with the ironing, washing dishes and making food. Now, when I think of my own boys, I am amazed at what I did. I wouldn't let my children near a toaster, never mind an iron! The mere idea seems ridiculous. Yet that was life and we just got on with it.

I wasn't resentful in any way for what some would see as a 'lost innocence' or 'lost childhood'. Such ideas are something we learn as adults. Far from it. I had plenty of time for me, plenty of friends and so much freedom to play and to discover the world. In fact, I had far more freedom than my own kids have. We had to bring ourselves up to some extent. We lived outdoors in the fields and on the street. We made up our own games and fun. Kenneth and I would say goodbye to Mum in the morning at weekends and turn up in the evening having played all day in the fields making dens. We regularly got lost in the woods and had to be rescued by one of the local dads as the light was fading in the evening.

Mum ensured that, despite everything, we still had a magical childhood. In fact, her ability to carve something beautiful from the pretty horrendous hand she had been dealt is what marked me during my childhood. She embodied triumph in the face of adversity and hope beyond all reason. All our dreams could have died with my dad's accident and the fire. He was the breadwinner and in many respects the 'life and soul' of the household. He was a larger-than-life character who loved to entertain. His sudden departure left a huge hole in our existence.

Mum could have retreated into herself and been consumed by a bitter cycle of anger, fear and despair. If she had a lot to be angry about, she had even more reasons to be fearful for the future. How would she survive, feed and clothe four young children? Dad was the sole earner and she had no means of earning an income. Dad had no pension and very little in the way of savings.

Yet amidst this seemingly impossible situation she found the strength to do the only thing she could: protect and love us with all her might. In reality, she gave up her life for us. She had an incredible capacity to cope with the hand that she'd been dealt – the fear, the uncertainty and the exhaustion which must have overwhelmed her at times. She believed that somehow, however mysterious it might be, there was a plan for us. And that plan was one of love. For her these were not just words – it was a way of life.

Her faith instilled in me an almost unshakable belief that even in the face of seemingly impossible situations, there is one force that can still overcome: love. But hers was not any kind of love. What she really taught me was that the power of a mother's love is so strong she will risk life and limb to ensure her children have a future. It is a love that refuses to give up on hope, even though every logic would tell you that hope is dead.

Her love was rooted in a deep Christian faith. It was a living faith in the oneness of everything and a belief that beyond what we see, the mysterious 'providence' of the universe would quite literally sustain us. This faith soon became a way of life for us – a survival strategy. It consisted of a simple, yet quite astonishing, belief: ask and you will receive; give and there will be gifts for you.

Some of the stories of 'providence' in our family are the stuff of family legend. They always followed a certain pattern. We needed (or wanted) something badly. Mum had no money but certainly wasn't lacking in faith. She would get us to say a prayer of intercession together and, as if by magic, the desired object would appear. It seemed to work every time – or at least that's how I remember it.

Mum always tells the 'story of the cars'. We needed a car to get to school and at some point the old yellow Peugeot we had since Dad died gave up the ghost. Mum had no money for a new one so things started to get a bit tough. We had to take the bus to school and then walk up a big hill. After a few weeks of walking back and forth, Mum decided we needed to 'ask' for a new car. So she got us together and we said our prayer for help. The next day we walked down to Mass together. Someone in the church spotted us all and asked Mum how things were. She mentioned the car trouble and how we were walking everywhere. The next day, some local sisters arrived at the door and handed Mum a set of keys. They had a spare car they weren't using at that time and they told Mum she could use it until she sorted something out. We were delighted. Then an envelope arrived containing £500 from an elderly parishioner. She had been praying the Rosary and our family came into her mind – this was a gift for whatever Mum needed and was enough to buy a small car. The power of community and prayer sustained us. Whether it was cars, holidays or even Christmas hampers, our prayers were answered – Mum could write her own book of stories about providence!

Some of it rubbed off on me too. I'll never forget going to my first music lesson in secondary school. We all got to test out musical instruments and I seemed to have a bit of a knack for the flute. I came home all excited and asked Mum to buy me a flute so I could take lessons. She looked at me with great love and said, 'Well love, you know I don't have money for a flute. But if God wants you to learn he will surely find a way.' So we said our prayer and left it at that. That same evening my mum's cousin, who was a nun in Ireland, called my mum. She wanted to let me know that, in a remarkable coincidence, someone had left a flute into the convent that week and she had thought I might want to learn. She had put it in the post to me. I was astonished. A couple of days later a beautiful silver flute arrived in a big black shiny case. Believing in the providence of God, of the universe became a way of life for me.

Through the local church, together with the whole family, I became very involved in a Focolare youth group. Focolare is a global movement which originated in Italy in the mid-twentieth century. It involves people of all faiths and none working together in whatever ways they can to build a more united world based on living out Jesus' commandment to love one another. My mum embraced the community-centred Focolare spirituality and in many ways it became a big extended family for us.

At a time when most people didn't really travel abroad that much, being part of such a global movement also gave me huge opportunities and opened my mind to the big world out there. In 1987, at the very peak of my teenage quest to save the world, my dreams came true when I was asked to represent the youth of Focolare at a meeting of world religious leaders in Kyoto, Japan. The meeting of the World Council for Religions for Peace was only the second time ever that such prominent religious leaders had come together. I was invited to read the declaration for peace signed by almost two hundred thousand children from around the world. It was a huge honour and the most amazing adventure for a fifteen-year-old girl from Falkirk!

I still remember Mum waving me off at Falkirk High railway station as I boarded the London train on the first leg of my journey to Japan. Goodness knows what was going through her head – but mine was buzzing with excitement. I met my travel companion Helen in London and together we took the flight to Tokyo via Rome, Denmark and Alaska. In Japan we met up with another ten children chosen from around the world to attend the meeting.

The week spent in Japan was pure magic, yet tinged with growing awakening to the inhumanity of war. I will never forget a solemn visit to Hiroshima on the anniversary of the nuclear bomb, where we prayed before the steps of a building shell. All that remained of those killed was an imprint of their shadows on the stone.

One moment of that week still stands out for me. After the inter-faith meeting where many important speeches were made and Helen and I got to deliver the children's message for peace, the whole group travelled to the holy shrine of Mount Hiei, a mountain north-east of Kyoto, for a Buddhist prayer.

The misty rain had started to descend as the faith leaders gathered on the stage for their final prayer. I still remember watching them parade by in their colourful garb: bright oranges, reds, greens, whites. Once all the leaders were assembled, the Tendai Buddhist leader, Rev. Etai Yamada, then aged ninety-seven, unexpectedly called the twelve children in attendance to the front of the stage. There were Christians, Buddhists, Sikhs, Muslims, Jews, Animists (those who believe objects, places and creatures all have living souls), all in different coloured costumes, all praying in their own way. There, in front of everyone, he gave the most inspiring speech. He said that when it rains, the Buddha is crying tears of joy. He thanked each of us for coming and said that he dreamed of a better future – one he would never see. He looked at us and told us he saw the hope of one in us and was placing his faith in us to build it. Together all the faith leaders prayed over us, each in their own way, that we would be able to live up to that dream. That blessing, in so many ways, set the course of my life.

A WAKE-UP CALL _____

My wake-up call to the state of the world, and particularly climate change, came early in life. I pursued it with all the zeal of a youthful idealist. It was underpinned by a deep desire to make things better. It was inspired by my mum's unshakable faith that no matter what life throws at us, we are each called to make a unique, positive difference. I got involved in countless social and environmental campaigns and started my own. I spent nearly a decade travelling across the world, completing a PhD in Economic Geography and then working on poverty and inequality. I was always questioning why our world is in the state it is in. I wrote and edited several books on the links between economics, faith and social injustice.

However, quite a few years passed before I realised how serious the environmental situation had really become. Climate issues had been pushed to the back of my mind. Perhaps, like most young people in the 1990s, I was taken in by the belief that governments had the situation under control. Those were optimistic times marked by the end of the Cold War and a sense that the West, with its democratic values, standards and human rights, had triumphed over dictators. The dawn of a new millennium seemed to beckon in a new age of peace, prosperity and harmony for all.

We had grounds to be optimistic too. Governments were working together more than ever before to sort out environmental issues. After the Rio Earth Summit in 1992, world leaders had managed to agree to fix the hole in the ozone layer and signed the Kyoto Protocol on climate change. It seemed that the world was on the cusp of a bright new age.

In late 2004, however, I had a rude wake-up call to the reality of climate change. I was lead researcher for a future studies project called 'Leading Edge' for Trócaire. The project was designed to help shape the direction that international development organisations would take for the next decade. It was basically a series of in-depth interviews with leading academics, policy makers, politicians and other intellectuals, about the future of the world. It was an amazing opportunity to ask some very intelligent people what we needed to be thinking about and doing in the future. It was like gazing into a crystal ball.

As I went around the world interviewing almost one hundred people, a very disturbing pattern emerged. All of them, without exception, said that the biggest issue facing poor people, and indeed all of us, in the next decade would be climate change. This came as a real surprise to me and my colleagues. Very few of those involved in international aid at that point in Ireland and internationally were talking about climate change as a risk to poor communities in the developing world. If it was being talked about, it had little or no bearing on what organisations actually did.

People who worked on global poverty back then tended to think that climate change was an issue that the likes of

Greenpeace and Friends of the Earth were taking care of. It wasn't our problem. It had more to do with polar bears and penguins than solving the important issues of human poverty and human rights. In fact, there was an assumption that solving poverty in developing countries was actually impeded by focusing on environmental issues. What business did well-meaning white people have telling Amazonian leaders they couldn't use the riches under their forest? Who were they to suggest putting the brakes on economic growth of the poor countries so as to protect the oceans?

The role of those working to solve poverty was to help people to attain their human rights. The idea that climate change could also have an impact on those rights – and that we needed to think about it – was actually something quite new. I remember at the end of the project I had to make a major presentation on the findings to all the staff of Trócaire. As I stood up to speak in front of a crowded hall, I felt quite uncomfortable. These findings didn't fit neatly with our plans. They could be easily misunderstood. They could shake things up. The evidence that had emerged from the Leading Edge project presented a clear choice: we could either ignore what was said or we could act on it.

Trócaire took up the gauntlet that had been thrown down by the Leading Edge report and made a decision to put the impact of climate change at the forefront of its work. That study was the start of a ten year (so far) campaign in Trócaire to bring this issue to the forefront of political life in Ireland and internationally. It has given rise to numerous projects in Africa, Asia and Latin America to support people in developing countries to cope with the impact of climate change and to make better plans to adapt to it. Moreover, the same project led to the start of the 'Stop Climate Chaos' coalition in Ireland.[1] Environmental groups and global poverty organisations joined forces to campaign against Ireland's poor performance on climate change. It is a campaign that has had a significant impact and is still growing stronger.

The Leading Edge project helped to deepen my understanding of climate change and the various ways it is affecting poverty; however, even having done all this research, looking back now, I can see that I didn't make the links to my own life and my own behaviour. I regarded climate change as something happening 'over there' – a big global issue which I could talk about passionately from the podium, but with few immediate consequence for my personal life or choices. It was my job, something I did from nine to five. I cared about my work deeply, but drew a line between what I did in my professional life and everything else.

This sense of compartmentalising my life started to change when I became a mum. This may have been a coincidence, but I think it was something more than that. How I thought about things definitely changed. After I had my own children, it was like those hard statistics about poverty, inequality, environmental destruction and so on, gradually took on new meaning. My emotional resistance to what lay behind each number got eroded somehow and I started to see those numbers for what they actually were – people's lives, families like my own. This became especially true when it came to climate change predictions. What were previously dry mathematical models and time series analysis took on a whole new meaning – one that was literally filled with children. And there, among all the children, were my own two boys.

One particular moment sticks out in my mind. It came about through a fellow Scot and friend, who also happens to be Ireland's leading climate scientist. Professor John Sweeney was one of the many scientists involved in the UN sponsored Intergovernmental Panel on Climate Change, whose work was recognised in 2007 with the awarding of the Nobel Peace Prize. Like me, John comes from Glasgow (in fact in a bizarre co-incidence we discovered we were born in the same tiny maternity clinic in Govan) and studied Geography at the University of Glasgow, my *alma mater*, before moving to Ireland.

On a fine Spring morning a few years ago, not long after I had returned to work after two spells of maternity leave, John and I travelled together to the Northern Ireland Assembly at Stormont to launch the Trócaire *Feeling the Heat* climate change report together.[2] This report came about through collaboration between Ireland's climate change institute, ICARUS, and Trócaire. It brought together the best available science of the fifth Intergovernmental Panel on Climate Change's (IPCC) assessment report on climate change,[3] with Trócaire's experience of developing countries. These periodic UN reports, which have been produced since 1990, provide a state-of-the art global assessment of climate change involving thousands of scientists. Our report was the result of a year of work between Maynooth University scientists and Trócaire.

We settled down in the gallery of Stormont and John started to give his presentation on climate change. With the dispassionate objectivity of his scientific profession, he started to explain the first consequences of climate change in Ireland and internationally. It is a familiar litany of tragic events. More floods, droughts, erratic weather patterns, exceptionally high temperatures, and so on. He then went on to describe some of the more human effects – communities forced to migrate due to failed crops, the appearance of new diseases, the salination of fresh water, the disappearance of many species of animals and plants. It was a grim picture, a dystopia, but one I was quite familiar with. Nothing about climate change is pretty.

He then went on to explain some really important scientific facts around climate change, underlining why it is happening and the time frames involved for solving it. As he went through all of the important facts, which I had heard many times before, something shifted in my brain. I understood something I hadn't quite grasped before. I understood that the air we breathe in and out every second and take for granted, is made up of billions, trillions of tiny molecules. These are all held in a fine balance. Over time, the composition of the atmosphere has changed due to major events on earth and the earth has responded with changes in the climate.

Yet in the last two hundred years, something profound has happened to tip the balance of the air we breathe. As a result of industrialisation, based on the burning of fossil fuels and intensive agriculture, we have released a phenomenal amount of carbon dioxide (CO_2) and other trace gases such as methane into the atmosphere. Fossil fuels such as coal, oil and gas are basically ancient stores of carbon from a previous age which are buried below the earth's surface. By burning them, we were able to achieve so much good. We advanced our economies, sped up production and gained so many benefits in terms of comfort, life expectancy and so on. But what we didn't realise, until relatively recently, is that these same sources of energy have a massive down side.

When we burn those fuels they release CO_2 into the atmosphere. This CO_2 doesn't go anywhere. It gets trapped in the lower atmosphere for a long, long time and cannot escape into space. It just sits there, mixing with the air around it. It then starts to act as a shield which allows heat from the sun into the atmosphere, but it cannot escape fully. As a result, the earth starts to heat up. Since the start of the Industrial Revolution, this process has accelerated as economies have grown, populations have risen and we have continued to release more and more greenhouse gases. The problem, which I hadn't then grasped, is that about half of the emissions from every machine, every steam engine, every woollen mill, every car, every plane, and every fossil fuel ever burnt since the Industrial Revolution are still up there in the atmosphere. So, the emissions from the dirty old van my grandfather drove during the Great Depression and the coal-fired steam train my great-grandmother took during the Second World War are still there, still having an effect on us today. To remove all the human-emitted CO_2 from the atmosphere by natural processes will take hundreds of thousands of years.

John explained that if more and more CO_2 is added to the atmosphere, the cumulative effect is to warm the atmosphere.

Depending on the amount of CO_2, the temperature will rise by 2, 3 or even 4 degrees. That doesn't sound like an awful lot, but it takes a huge amount of energy to heat up a whole planet! And it is not true to say one degree doesn't matter. The difference between zero degrees and one degree can be very significant – especially when it comes to a block of ice or boiling water. Ask any child who tries to eat a melting ice cream. Or anyone who has forgotten about the potatoes in the steamer and returns to find a burnt pot. Or someone late for work trying to scrape thick ice from their windscreen. Nature doesn't work in gradual lines but in critical thresholds that change states from solid to liquid, liquid to gas. So, as a result of emissions due to existing pollution, we have actually reached a global temperature rise of $1°C$ already. That's a heck of a lot of melting ice around the world.

As I sat listening, a simple image came into my head. I thought about a balloon game I play with the children at birthday parties. I blow the balloon up as much as I can, until I am red-faced, to see if I can make it explode. It causes great excitement as the kids scream and jump around. Well, pumping CO_2 into the earth's atmosphere is a bit like that. At first the air in the balloon doesn't seem to make much difference. It inflates slowly. Then, when it reaches a nice comfortable amount, we usually stop blowing and tie the knot. However, there is only a certain amount of air you can put in before it is full up. If you keep pumping air into a balloon even when it is full, eventually the result will be dramatic. In the case of the balloon, the pressure will cause it to burst. You will probably have screams and maybe a few tears! In the case of the earth, if we keep pumping CO_2 in when the atmosphere is 'full up', or has reached its safe limit (its 'carbon budget'), it will overheat and become critically unbalanced. It will be catastrophic.

Back at the start of the Industrial Revolution, before coal was being used as fuel as a matter of course, the carbon balloon was deflated and CO_2 levels in the atmosphere were very low. Since the mid-eighteenth century however, the balloon has been inflating and inflating at an accelerating pace and there is no way to let gas out of the balloon even if we wanted to. When we understood the greenhouse effect back in the 1980s, the balloon was already half full but there was still a bit of space to play around with. We could still have chosen to inflate it more slowly or to invest in measures to absorb the CO_2. Instead, we blew harder. Now the amount of remaining space to blow new air into the balloon without effectively popping it is extremely small.

At this point in history, we need to urgently stop blowing air into the balloon. Instead, we are doing the opposite. Our lives practically revolve around pumping more and more air into it – but we have to stop!

Scientists, moreover, have put actual numbers on the space left in the balloon. According to the 'Fifth Assessment Report' the consensus is as follows: to have a reasonable chance of keeping the world within relatively safe limits of a temperature rise of under 2°C on pre-industrial times, we need to keep the amount of greenhouse gases released into the atmosphere by humans below the equivalent of 2,900 giga-tonnes of CO_2.[4] All the human activities since the Industrial

Revolution have already pumped 2,144 $GtCO_2$-e ('gigatonnes of equivalent carbon dioxide') into the atmosphere. If we do the simple subtraction (2900 – 2144) you arrive at the amount of CO_2 we can still emit: 756 $GtCO_2$-e. Given current trends, at present rates, this limit will be breached in under twenty years.[5] We will cross that safe limit by 2035.

Moreover, the world today seems to be in deep denial about these facts. We currently have enough oil and gas reserves of the top ten oil companies in the world to send the world into runaway climate change of 7 or 8°C.[6] Despite this, governments continue to encourage companies to look for more oil – even in the Arctic Circle – and are in deep denial about how this is fuelling climate change.

As I sat listening to John explain all this in Stormont, I suddenly felt overwhelmed with emotion. I found myself about to address a group of politicians and fighting back tears. Those dry facts which I had heard before suddenly became something else: my own flesh and blood. I fast forwarded those timelines in my head and suddenly saw my two little boys grown up. They were in their mid-twenties and setting out in life. It suddenly dawned on me that the great future I had imagined for them is under threat and may not come to pass. In fact, if what John was saying was true – and I had come to accept it as the truth based on the overwhelming scientific evidence – their future will be a lot more uncertain, a lot more troublesome than you or I would like to believe.

Moreover, for those on the front line of climate change, this is not a future problem. It is happening right now. As I got up to speak that day, I told the stories of countless families around the world who are suffering right now because of those CO_2 emissions – emissions they certainly never had anything to do with but which are already causing so much pain and destruction. Despite my calm appearance, however, inside I felt heartsick at the thought that the bubble of a happy future I envisioned for my children had been burst. When I understood the implications of John's talk that day, that feeling never left me. The science affected me in a deep way. It went from being theory that was purely academic to being something I internalised in my heart and soul. It instilled in me a sense of urgency about the changes we need to make – if for no other reason, for the sake of every child on this planet. It is their future that is hanging in the balance.

NOTES

1. *Stop Climate Chaos*, www.stopclimatechaos.ie
2. Trócaire, *Feeling the Heat*, 2015, www.trocaire.org/sites/default/files/resources/policy/feeling-the-heat-2015-1.pdf
3. Fifth Assessment Report (AR5), www.ipcc.ch/report/ar5/
4. Because different gases contribute to climate change, this equivalent value is used to capture the impact of all greenhouse gases. Based on UNFCCC, *Adoption of the Paris Agreement: Draft Decision/CP.21*, United Nations Framework Convention on Climate Change, 2015, unfccc.int/resource/docs/2015/cop21/eng/l09.pdf
5. Based on IPCC 'Fifth Assessment Report' calculations. The real-time carbon clock is available on the *Guardian* website, www.theguardian.com/environment/datablog/2017/jan/19/carbon-countdown-clock-how-much-of-the-worlds-carbon-budget-have-we-spent
6. Bill McKibben, 'Global Warming's Terrifying New Math', *Rolling Stone*, 19 July 2012, www.rollingstone.com/politics/news/global-warmings-terrifying-new-math-20120719

A HEAVY HEART _____

Absorbing this new understanding about climate change and what it means for our children was not easy. It is still not easy to be honest. I really wish it wasn't true. If I could click my Dorothy heels together and be back in Kansas, life would be quite different! My new understanding of climate change filled me with a deep sadness, frustration and even anger. These were emotions I wanted to push away at first, but I just couldn't switch them off. I couldn't look away.

Sometimes, coming to terms with this new understanding of climate change led to a growing sense of anxiety when engaging in even the most routine tasks, like taking the kids to school in the car or going on holiday abroad. This anxiety started to manifest itself more as I noticed abnormality in the seasons or subtle changes in the way plants in the garden are growing (or failing to grow). Above all, it showed itself in the growing sense of anxiety and helplessness about what I could do.

One day I was driving on the motorway and as I watched all the cars rushing by, I became aware of each of us alone in our own little bubble, gushing out CO_2, oblivious to the consequences of our actions. Some, like me, knew something about climate change but were stuck in an ethical quagmire. It was all very well to care about the climate, but stuff still had to get done *today* and ready-made alternatives don't always exist. As I drove, I could see in my mind's eye people whose lives had been turned upside down by rising seas. People living on tiny Pacific islands, their livelihoods in ruins, the seas encroaching on their land. I could see the women in villages across Africa walking miles for firewood to cook for their families with no prospect of gaining access to the cheap energy we have all taken for granted. And yet I kept on driving. I had to be somewhere fast.

At other times the country would be basking in a summer heatwave, with the weather presenters talking about how records continue to be broken. As I sat in my garden enjoying the sunshine and allowing my mind to be temporarily transported to the South of France, I would feel a sudden pang of anxiety – a deep guilt about the inevitability of the increasingly weird weather we seem to be experiencing. Those weather statistics delivered in such matter of fact, almost jolly tones – 'Scientists confirm this is the fourteenth consecutive warmest month on record – meanwhile Ireland basks in summer sun.' 'June 2016 was the warmest month on record.' – seemed to ring in my ears like deafening alarms.

Two facets of life – the innocent pleasure of an ice cream and the tragedy of a silent, unfolding crisis – were sitting very uneasily together in my head.

I realised that until that day in Stormont I had actually not *accepted* the profound truth of the science behind climate change, not really. I knew about the issues and I could sound clever talking about them, but I didn't actually believe deep down that this was true. Or at least I didn't let the consequences of that belief into my head. Not in the same way I thought other things were true. We know that gravity is true because we know that it keeps us stuck to the earth. We know that matter exists in different states – solid, liquid and gas – and that if you change the conditions, you can change the state of the material. We accept that all living creatures are born, live and die. We accept that the earth is round. We accept lots of things are true as a result of scientific enquiry. We internalise them into our minds and develop a kind of brain-muscle memory for them. You don't have to keep telling yourself they are true over and over again.

We accept these things as established scientific facts. Once these facts are established as *true*, no amount of our arguing about them can change what they are. That's why we call them 'objective'. As facts about the world, they exist outside of us, independently of us, but we also internalise them through education and through life experience. We verify them through our scientific method but we don't make them be. We accept them and build our lives around the intricate and beautiful mathematical formulations which underpin them. The problem with climate change is that despite the overwhelming evidence, for various reasons, most of us have not really accepted it as fact – at least not in the same way that we accept other facts like gravity!

Let's take humans out of it for a minute. Science and reason has brought us to an understanding of the physical world around us. The scientific enquiry enables us to observe and predict changes in the world's atmosphere based on objective, mathematical calculations of levels of certain gases. First, we speculated, then we enquired more, then observed, to the point that we now *know* beyond reasonable doubt what happens when CO_2 and other greenhouse gas levels rise in the atmosphere. They change the carbon cycle, which has an impact on the water cycle and consequently has an impact on weather patterns, the biosphere and on all ecosystems to greater or lesser extent. These processes, as far as we know, are not reversible. Once an ecosystem is changed, it is effectively lost, it is difficult to recover in all its diversity. This is not revolutionary science. This is basic science.

It isn't an 'issue', much less a 'policy issue' – it is essential information about new physical contexts in which we as a human species are living. It can be likened to the major transformative scientific discoveries in the past, such as the ancient discovery of the earth being round or Copernicus' understanding that the earth revolves round the sun. It is the physical context in which we 'live and move and have our being' (Acts 17:28). It can't be negotiated with. You can't talk to the climate and say 'please cool down' or 'please wait until the next election is over before sending us the next big storm'.

Nature cannot be fooled by what are seen as necessities of convenience or the political economy. Once we know the earth is round, no amount of arguing will make it flat. Nature cannot 'adapt' to accommodate the needs of our political economy, no matter how important they are perceived to be – adaption can almost always only be in one direction. In fact, as a very minor part of this planet, within the known universe, humans are subsumed within the whole and regardless of what we like to think, won't win any 'fight' against climate change. If we put ourselves on a collision course with the planet, there will only be one winner.

There is a kind of panic which results from grasping this reality; hence, the urgency of tackling climate change. This was really brought home to me during an encounter with another climate scientist called Professor Kevin Anderson, an eminent scientist who, like John, has a rare talent for communicating the issue of climate change in language that everyone can understand. I met him at a seminar in Maynooth University when he was reviewing the Paris Climate Agreement. He wasn't too positive about the capacity of humankind to come up with the kind of solutions we need in the time frames left. The Paris Climate Agreement, he said, is a good political agreement. Those of us who were present in Paris during the talks actually thought it was a miracle to have a consensus of any kind. The problem is that while it is a miracle of diplomacy, in terms of what the science tells us needs to be done, it doesn't go far enough.

Professor Anderson made the point that despite this agreement, and all the agreements that have gone before it, the emissions pathway (the direction we are heading in) has not changed even remotely. Even if all the commitments in the Paris Agreement were honoured, the world would still be heading far beyond what scientists agree is a 'safe' temperature threshold.

The problem, according to Professor Anderson, is that we have been here before in terms of treaties, promises, false starts. Very little seems to be budging the collective will of most governments to really recognise what is at stake here. Given how much humanity has delayed in making the necessary changes, moreover, only one decade now remains to take action. If you look at the emissions trajectory, it still points upwards. It is relentlessly stubborn. We, as a species, have ten years to bend that trajectory into an arc, and eventually to reduce our emissions to zero.

The short time frame now available is really key. It means that the easier options to reduce emissions that existed in the past are now less relevant. If we had started in earnest ten or twenty years ago, we could have done many things to reduce our emissions. We could have invested in inventing carbon capture technologies; we could have embarked on the most ambitious global campaigns of reforestation and so on. But given that we only have one decade left and many of these actions require a much longer time frame – and prohibitive amounts of money – they are simply not feasible solutions any longer. Some projects under consideration, such as investing huge amounts of money in geoengineering strategies involving creating mirrors in space, spraying aerosols into the upper atmosphere to deflect the sun or seeding sea water into clouds are actually dangerous distractions – ridiculous gambles on our children's future. They may also have many unintended consequences.

Others are just too impractical to be seen as the main solution. In fact, Professor Anderson ended his talk in Maynooth by saying (half-jokingly) that the big invention we really need now as a species is a time machine. If we had that we could go back a couple of decades, or even back to the start of the Industrial Revolution and undo the damage. While such talk may seem defeatist, it highlights the dire situation we now are now in. The only solution, he says, is a radical shift in the production and consumption patterns of the wealthiest populations – the ones who are emitting the most.

The scale and pace of change Professor Anderson seems to be proposing is breathtaking. The chances of success seem low. As one comes to terms with this new understanding, it can lead to defeatism. What if all this campaigning is actually in vain? This acute sense of time slipping away, and the need to make actual, real sacrifices, is a dilemma faced by anyone passionate about a big cause. There is a risk that in all the activism to protect their future, I could miss my children's present, especially the precious years of childhood. There is no guarantee that all this effort and time will do anything other than make me exhausted. Why not shut the door, batten down the hatches and hold on to what we have.

Taking climate change seriously can turn your life upside down. It is no surprise then that psychiatrists are pointing to the mental health effects that climate change is already having – particularly on the young. At times the reality of climate change can be overwhelming. In those moments, I have to keep my eyes firmly fixed on the reasons such a struggle is necessary and worthwhile. It takes all my reserves of hope and that belief instilled in me since childhood: there is no force more powerful than a love that knows no bounds. It is a time to dig deep and remember the roar of the lioness who would stop at nothing to protect her cubs.

ONE PROBLEM, MANY SOLUTIONS __

While the scale and urgency of climate change can seem quite overwhelming, the main cause of the problem itself is actually quite simple. We often tend to assume that big problems are inherently complicated ones and we shy away from complicated matters. But in the case of climate change, the cause is actually quite straight forward. The basic science around it has not changed since I did that project in secondary school in the 1980s. In fact, the science around it has been known for almost two hundred years, since the Irish physicist John Tyndall first understood the processes that keep the earth warm. The science of climate change, while it has developed exponentially in recent decades largely due to the power of computer modelling, points to the fact that certain human activities, especially burning fossil fuels and over-dependence on industrialised agriculture, are causing the earth's atmosphere to warm up. That's a bad thing if you care about the kind of world our children and their children will grow up in.

Grasping the simplicity of the problem is really important when it comes to thinking through what needs to be done and where to focus energy. The really good news is that the fact the problem is quite straightforward means that there are a few big solutions which can make a dramatic difference on a global level. In the second half of this book I will explore what we as individuals and communities need to be doing and how this connects into the big change. Before coming to that, however, I want to set out the big picture of what the scientists tell us needs to be done globally.

The first thing we need to do is to stop – or at least dramatically slow down – the amount of damaging CO_2 we are pumping into the atmosphere. If the imaginary balloon is nearly full, as I explained before, common sense suggests that we all need to take urgent action to stop it 'exploding'. That's an absolute given. It has to be the first response. How many times have I heard my kids shout, 'Stop, Mummy, stop! You are going to burst it!' Once the balloon explodes, or global warming goes beyond a certain level, it will be impossible to reverse the situation. Many of the impacts of climate change will be irreversible.

Slowing down the amount of CO_2 and other gases getting into the atmosphere is what climate scientists call 'mitigation'. The dictionary definition of mitigation is, 'The action of reducing the severity, seriousness, or painfulness of something.' In the case of climate change, mitigation is another word for slowing down and eventually stopping *future greenhouse gas emissions*.

By far the biggest source of pollution comes from burning fossil fuels such as oil and gas, which are currently the main source of energy for industrial processes, for electricity, for home heating and for transport right across the world. Offsetting the damage could seem almost impossible as these fuels drive the global economy.

Yet there are reasons to be hopeful. The great news is that there are many alternatives to burning fossil fuels now available. It is possible to talk about an 'energy transition' to clean energy. In the last few years, the amount of energy available from wind and solar power has jumped dramatically. These are all clean, renewable energy sources. Solar energy availability jumped by 50 per cent globally in 2016 alone.[1] This is truly remarkable and is already transforming our ability to continue to produce things and heat our homes. Every week there seems to be a new report about the falling cost of producing solar energy.[2] Business people are talking about a 'rooftop revolution' taking place. Entire countries, such as Nicaragua, Sweden and Scotland are now vying to become the first 100 per cent renewable energy countries.

More and more, there is a shift to make these renewable energy schemes more community friendly. Whereas in the past there was an emphasis on large grids where energy was owned by a few and distributed to everyone else, in many European countries there is a shift now towards making renewable energy more democratic and local. Communities have come together to take ownership of their own energy sources and get a stake in the local turbines or solar panels. In some countries people who put solar panels on their homes can feed into the energy grid and benefit from the energy they save at home (especially given that the sun usually shines when people are at work) by selling it back to the energy supplier. It is a system which has been working beautifully in Germany and in many other EU countries for over twenty years.

This energy revolution is also really positive for people living in poorer countries and communities. It potentially removes the need for expensive central government investments in power plants and electricity grids. Now, with the dawn of cheap renewable energy, many poor communities can 'leapfrog' this phase of slow, dirty development.[3] Countries are finding that they are able to become energy sufficient and secure, providing all the benefits of electricity at a fraction of the cost of the past. It is leading to a real energy revolution supported by a new generation of development workers.[4] The SHINE campaign, launched in 2016, aims to help accelerate the availability of cheap, clean and local energy across Africa.[5] The clean energy transition is already under way, but needs to go much faster if we are to stay within safe climate change limits.

Fossil fuels, of course, are not just used to generate electricity. They are also the main fuel used in most forms of transportation. So, shifting to low carbon modes of transport is also essential to lowering the rate of emissions. The vast majority of vehicles which run on petrol and diesel are among the worst culprits for pumping CO_2 into the atmosphere. The good news is that alternatives to these vehicles now exist and are becoming increasingly affordable. Of course, an electric vehicle is only of benefit if there is also the means to produce electricity without pumping CO_2 into the atmosphere. Where that exists, electric vehicles represent a major step forward. Shifting transport systems to these types of vehicles – whether cars, buses or trucks – is a significant way to mitigate against rising levels of CO_2. Of course, there are still climate emissions involved in the manufacture of such vehicles – but the transition to ridding the world of combustible engines is essential. Thankfully, it is already happening and is gathering pace. Many countries are now considering banning combustible engines in cars in the next twenty or thirty years.

In terms of transport, however, there are much simpler climate-friendly alternatives than getting into cars which need to be part of the solution to climate emissions. Many countries are now focusing on how cities, in particular, can be adapted to encourage walking and cycling for short journeys rather than using cars and buses. Most of these journeys tend to be made on a daily basis – for example, taking children to school – so making them 'zero carbon' journeys would make a significant contribution. It would have the added benefit of a healthier populace too!

For some means of transport, of course, it is hard to envisage readily available alternatives in the short-term. Tackling emissions from aviation is so difficult, it has, by and large, been kept off the agenda of international climate negotiations. The problem here is that many people's lives in a globalised world (particularly the wealthier and more influential) have become structured around cheap air travel, whether for work, holidays or trips abroad to see family. The contribution of the aviation industry to climate change is the equivalent of that of an industrialised nation like Germany. Flying has a seriously detrimental effect on the climate, as well as on the environment in general. It is estimated that on average each passenger on a flight emits the equivalent of a four-seater car per kilometre of journey.[6] When you think of the number of people on each flight, and the long distances travelled, the scale of the issue becomes clear. Each year over three billion flights are made – just over eight million per day. Airplanes may be becoming more fuel-efficient, but nonetheless, there is no way to transport people across continents quickly that doesn't involve the use of vast quantities of fossil fuels.

Yet the reality is that many millions of people, including myself, have built our lives around being able to travel. Our world has become globalised and our families and professional lives are interwoven across continents. Dealing with this is a very tricky issue which I will come to later.

Another significant source of climate emissions, which is very well documented, but perhaps less widely known, arises from food production and consumption. Our globalised food system is very sophisticated in some ways – mass-producing food which can be transported, stored and consumed practically anywhere in the world. However, this same system is extremely inefficient in other ways, particularly when it comes to climate change and broader environmental issues around resource depletion and the destruction of bio-diversity. Vast amounts of emissions come from the change of land use from forests to arable land, industrial agricultural processes, transportation of food over large distances, and industrial-scale processing and packaging of food. At the end of the food chain, then, vast amounts of this highly carbon-intensive food has to be destroyed as food waste.

Other food-related emissions come from the particular products the global population likes to eat. Across the world, there is growing demand for red meat and dairy products, especially as the middle classes in big populations such as China and India grow.

The contribution that the production of beef and dairy products play in global greenhouse gas emissions, and especially those of countries like Ireland and New Zealand with high numbers of cattle relative to people, is significant.[7] Unlike most other farm animals, such as chicken and pigs, cattle release large amounts of methane into the atmosphere during their lifetime. While significant efforts are being made in developing more carbon-efficient practices, in reality there is very little that can be done to resolve this issue as it is a result of basic biological processes. Methane, like CO_2, is a greenhouse gas and contributes towards warming the earth. In fact, methane is far more potent than CO_2. It stays in the atmosphere for a shorter amount of time but it has an impact on the climate which is over thirty times that of CO_2.

Moreover, red meat is not a very efficient source of food when it comes to converting inputs into outputs as calories. Red meat production is by far the most resource-intensive of any food type and has the most damaging impact on the environment. Red meat requires twenty-eight times more land to produce than pork or chicken and eleven times more water, resulting in five times more climate-warming emissions. In 2014, a study reported that when compared with staples like potatoes, wheat and rice, the impact of beef production per calorie requires one hundred and sixty times more land and produces eleven times more greenhouse gases.[8]

Clear alternatives exist in terms of a transition to clean energy, transport and local, plant-based food, which could help mitigate climate change emissions. Logic would suggest that pursuing these alternatives should be a top priority for governments when it comes to mitigation against climate emissions. Among other things, it would involve governments throughout the world examining their laws, policies and tax systems to encourage a transition towards the existing alternatives – and support for those working in economic sectors which need to change.

At the same time, keeping in mind the metaphorical balloon growing dangerously bigger, logic would dictate that we need to do everything we can to reduce all emissions right now. The alternatives outlined above will take years, if not decades, to fully implement. In the meantime we should be using remaining 'stores' of greenhouse gases sparingly. Imagine if you were caught in a snowstorm and couldn't get out the house. You had very little fuel left to keep the house warm. Your first instinct would be to economise – use it as sparingly as possible. You certainly wouldn't be frittering it away by leaving the thermostat up to maximum, with windows and doors open. The same applies to seven hundred and fifty-six gigatonnes of greenhouse gases that remain.

As we make the transition to a 'new' energy, food and transport system, we also need to minimise the impact of our 'old' habits. We need to make these old ways of doing things 'greener'. This is where technology companies have been focusing their attention in recent years. Take the example of buildings. It would be nice to go and build new eco-friendly, zero-emissions buildings for everyone, but that isn't possible (and also would be inefficient). Most of us have to make do living and working in older buildings which were constructed long before climate change was taken seriously.

Buildings account for a substantial amount of global greenhouse gas emissions. Some of those emissions come from the construction of the buildings. Others come from the inefficiency of the buildings themselves. Just think about how much energy is wasted in homes and buildings right across the world. Many things can be done at a national and international level to reduce this energy waste. New buildings can be built to much higher standards. Older ones can undergo deep retrofitting. If all our public buildings and private homes across the world were retrofitted, it would result in a significant decrease in our CO_2 emissions, as well as lower heating and energy bills.

The same logic applies to transport and food – if we have to travel by car, make sure it is the most fuel efficient car possible. If we have to travel by plane, make sure the plane is as efficient as possible. If we really must eat red meat, then make sure it comes from as sustainable a source as possible. All of these efficiency measures add up and are essential as part of a transition to a zero carbon world. They are captured in the growing shift within the corporate world towards a 'green economy', with a strong focus on measuring the sustainability of goods and services.

If governments across the world enabled these changes to happen, it would make a significant dent in global emissions. In fact, this small number of key shifts would go a long way towards tackling the climate problem. Many of the options which I have discussed above involve a serious financial investment in order to see a significant reduction in emissions. It is just about possible, however, to see how they might be possible if governments aligned their policies and people became more engaged about what is happening to the world.

Such measures are just the start. Many leading climate scientists are now saying that while all of the above measures are essential, given how full our carbon balloon is and how little time is left to address the problem, more radical measures are necessary if we are going to have a chance of avoiding runaway climate change.

This is where the final part of the mitigation jigsaw comes in. As well as seeking alternatives and becoming more efficient and green, we also need to *cap our emissions*. This is perhaps the most complex part of the mitigation equation because it strikes at the very heart of our capitalist economic system. Those of us who consume too much need to also be asking ourselves how much is 'sufficient'. We need to rapidly relearn the idea of 'enough'. In fact, our ever-growing emissions are not just like a leaky pipe we can fix – but symptomatic of a badly thought out economic and financial system which compels us to extract and use more and more.

This economic system is built around generating more and more unnecessary *wants*. Yet every single substance we extract from the earth, produce or move around incurs a cost in terms of emissions (and wider environmental damage). This is where the really difficult issues begin to arise in relation to climate emissions. Dealing with this systemic issue involves asking ourselves some pretty fundamental questions about our economies and our lifestyles, especially in Western societies. It means addressing the extreme inequality of emissions – the fact that over half of global emissions come from a minority of 10 per cent of people (including myself and most people reading this book).[9] Most of us are continuing to consume far beyond our planet's means with little regard for the consequences. It also means questioning the way in which other rapidly growing nations are choosing to develop, and whether they have a right to pollute in order to help their citizens overcome poverty and meet their basic needs. It is where the question of 'climate justice' – the rights of the poor and the rich to pollute and repair damage – come into sharp focus.

So-called 'sufficiency measures' are where the least progress is being made in tackling climate change and where urgent progress is now needed. In fact, all the data shows that even if we do things as efficiently as the available technology allows, such as save on energy, much of that efficiency gain is more than compensated for by having more money to spend on something else. Generally, that 'something else' generates more emissions. It is absorbed back into an economic system which has a toxic addiction to extracting precious resources from the earth and producing carbon emissions.

Sufficiency measures are hard to address because on the surface – and let's be clear, sometimes in reality – they involve those with money renouncing those things and experiences we associate with pleasure and a fulfilling life. But they also involve thinking about how we distinguish between those who are still in desperate need, and should have a basic right to consume more, versus our wants. Taking sufficiency seriously, for example, would involve shopping less – at least for brand new stuff (and dealing with the knock-on economic impacts). It would mean changing our tax system to ensure we pay an awful lot more for things that emit a lot of CO_2; for example, beef and flights.

In short, if we talk about sufficiency, the debate becomes a lot more about hard choices and what we think is needed for a happy and fulfilled life. These thorny issues are still a million miles away from most government agendas – and those of most citizens who elect them into power. And this is no accident. It is where the climate change question comes face to face with the key drivers of economic growth and political success or failure.

Putting in place the kinds of measures necessary to deal with sufficiency would require a radical shift in the national and global political system. All states rely on economic growth to generate tax returns to enable them to pay for public services – health, education, transport infrastructure, defence, welfare – that citizens demand. The more the economy is growing, the more people are in jobs, the more profits are made and the more the government is able to raise taxes. Governments want us to produce and consume more and more so there is a bigger economic pie to divide up. If there is less stuff, then where does the revenue come from? Where will future jobs come from?

These are huge questions which strike at the heart of progress and development. They are sparking a lot of new thinking around the necessary economic transition to address climate change. We need to be thinking a lot more about jobs in a 'sufficient' economy. Which jobs add value through eliminating waste or reducing consumption? Which jobs are needed to add to the lifespan of existing products? What jobs emerge in the spaces where people consume less – and have more time for other activities? There are many new jobs in a sufficient economy – but enabling them to thrive will also involve some pretty radical changes to our laws, taxation system and fiscal policies to incentivise climate action. Setting clear caps for emissions for different sectors of the economy is one important way this can be done.

A big political problem in implementing these more radical measures is the time frames in which most governments work. Electoral cycles can lead to avoidance or delays in implementing unpopular measures which might be in the long-term interest. On the face of it, some measures to cap carbon emissions would be hugely unpopular. Just imagine the public reaction to a high carbon tax being imposed on flights or on beef and dairy production! Such moves, if implemented in the absence of a proper plan to deal with the resulting upheaval would most likely not win many votes. This means that citizen education on the problems we face as a result of climate change becomes critically important.

Furthermore, many governments tend to be highly influenced by key sectors in the economy which have a stake in preventing a more transformative change – or ensuring change is in their favour. In the case of tackling climate change, the influence of the powerful fossil fuel lobby has been central to delaying action. The fossil fuel industry has the most to lose in fighting climate change in the short-term and will go to any lengths to delay change.

Estimates show that current fossil-fuel reserves contain around 2,795 $GtCO_2$-e of emissions.[10] This amount is three times the remaining global carbon budget of 721 $GtCO_2$-e – and five times the estimated 565 $GtCO_2$-e share of the budget allocated to the fossil fuel industry.[11] The leading fossil fuel companies alone, if they continue with their current plans, have the potential to explode our balloon and send the world into meltdown.

These companies are well aware of this. In fact they have known for decades. The result is that, like the tobacco industry in the 1990s which sought to undermine the science around lung cancer, the fossil fuel industry spends millions each year lobbying governments to disregard sound scientific research and delay action. In fact, the Paris Agreement, which was signed in 2015, makes no reference to fossil fuels as the main cause of climate change.

The reality is that much would need to change in our politics, legal and financial systems to address the problem of climate change. Many of our laws and policies were put in place by people who had little or no knowledge of climate change. These individuals may have been well meaning and democratically elected, but they were also ignorant of the science and the need to take it into account. Many still are ignorant of the devastating impact climate change will have within a relatively short space of time. Reversing bad decisions or preventing them from going ahead requires radical changes in policy in order to bring our legal and political system in line with what scientists are telling us. There is a lot of catching up to be done in legal and political fields to integrate the truth about climate change into policies as diverse as energy, transport, agriculture, manufacturing and tourism. Even social policies like housing, education, welfare and health will need to change.

When these powerful national and corporate interests collide with the international political processes to curb climate emissions, the result is generally a stalemate. The result is that the pace of change in achieving action on climate change has been far too slow. Governments have been talking since 1995 and emissions have continued to rise. The Paris Agreement was signed in 2015, but even this landmark treaty doesn't set out a clear pathway to restructuring the global economy on a path towards sufficiency. Instead, it essentially allows all countries to set their own emissions targets and then adds them all up. It throws the whole issue of tackling climate change back into the hands of individual governments, which are perhaps the least equipped to make change happen. Moreover, there are no real sanctions for not complying with the treaty.

Bill McKibben, the climate campaigner, wrote in an article for *Rolling Stone* in 2017 that 'winning [the battle with climate change] slowly is another way of losing'. Because of the nature of climate change, and its dependence on critical thresholds beyond human control, there is no time to lose. Given the urgency of scaling up and implementing solutions and the abject failure of governments to tackle the problem, the role of individuals, communities and other non-governmental organisations is becoming increasingly important. In fact, it is becoming vital. No other force at present is capable of making change happen. Key to this is understanding. It is about really opening our hearts and minds to what is at stake in the choices we make: the future of our own children and their lives on this planet.

NOTES

1. Adam Vaughan, 'Solar power growth leaps by 50% worldwide thanks to US and China', *The Guardian*, 7 March 2017, www.theguardian.com/environment/2017/mar/07/solar-power-growth-worldwideus-china-uk-europe
2. Robert Fares, 'The Price of Solar Is Declining to Unprecedented Lows', *Scientific American*, 27 August 2016, blogs.scientificamerican.com/plugged-in/the-price-of-solar-is-declining-to-unprecedented-lows/
3. Georges Alexandre Lenferna, 'How Africa could leapfrog fossil fuels to clean energy alternatives', *The Conversation*, 2 March 2016, www.theconversation.com/how-africa-could-leapfrog-fossil-fuels-to-clean-energyalternatives-55044
4. Wendell Roelf, 'Sun, wind and water: Africa's renewable energy set to soar by 2022', *Reuters*, 15 November 2017, www.reuters.com/article/us-africa-windpower/sun-wind-and-water-africasrenewable-energy-set-to-soar-by-2022-idUSKBN1DF1T8
5. 'SHINE – Investing in Energy Access For All', *The Sustainable Energy For All Forum*, www.seforallforum.org/session/shine-investing-energy-access-all
6. 'Average passenger aircraft emissions and energy consumption per passenger kilometre in Finland 2008', *LIPASTO – Calculation System*, VTT, 7 May 2009, www.lipasto.vtt.fi/yksikkopaastot/henkiloliikennee/ilmaliikennee/ilmae.htm
7. Damian Carrington, 'Eating less meat essential to curb climate change, says report', *The Guardian*, 3 December 2014, www.theguardian.com/environment/2014/dec/03/eating-less-meat-curb-climate-change
8. Gidon Eshel, Alon Shepon, Tamar Makov and Ron Milo, 'Land, irrigation water, greenhouse gas, and reactive nitrogen burdens of meat, eggs, and dairy production in the United States', *PNAS*, 19 August, 2014, 11 (33), 11996–12001, pnas.org/content/111/33/11996
9. Timothy Gore, 'Extreme Carbon Inequality: Why the Paris climate deal must put the poorest, lowest emitting and most vulnerable people first', *Oxfam International*, 2 December 2015, oxf.am/2FMYtY2
10. Refers to the fossil fuel reserves owned by companies, and countries that operate like fossil fuel companies.
11. As stated in the annual reports and strategic plans of the companies in question.

MORE THAN THE STATISTICS _____

As I started to understand what is happening to our world a bit more deeply, climate change started to acquire names and faces. Scratch below the surface of stories in the media, behind the mask shielding us from the consequences of our actions, and individual faces and stories start to emerge: families made homeless by extreme hurricanes; people killed in flash floods; homes turned to ash by wildfires; livelihoods decimated; lives destroyed. Over time, these few faces become many, to the point that climate change now has millions of names and faces. They are too many to count and their number keeps on growing by the day, the week, the month, the year.

Several faces of climate change stand out for me. In 2014 I made a work trip to the UN to attend a summit on climate change in New York. The former UN Secretary General Ban Ki-moon had called a heads of state summit to look at how the world was addressing climate change. There have been UN summits now for many years on this issue, and endless wrangling about how to reduce greenhouse gases. Those negotiations traditionally get mired in a bizarre race to the bottom and fighting over who is to blame and who should pay. This summit was to be different. It was called the 'ambition' summit. It was designed specifically to allow countries to go beyond their normal negotiating positions and to set out their most ambitious targets.

While I was there, I went with a Church delegation to visit a UN ambassador from one of the tiny Pacific islands already affected by a changing climate. We had arranged to escape the UN complex, which can be a bit of a circus during summits, and to meet the ambassador in their consulate. The first thing that struck me as we left the UN building, was the location of their embassy. Far from the shiny offices in Midtown, with fabulous views of the UN and the Manhattan skyline, her embassy was a tiny office on the Lower East Side, miles from where the main diplomatic action was taking place. We waited for her in the dark boardroom, staring at the only picture on the wall: a tiny circle of emerald in a turquoise ocean. A little bit of paradise.

The ambassador arrived and greeted us warmly. She was moved that we had taken the time to come to visit her – despite all the razzmatazz of diplomatic engagements on the sidelines of these big summits. After we had all greeted each other, she started to talk. She spoke with great pride and dignity, but it was evident from the first moment that she was angry. She was very angry.

Over the past few years her country had already been experiencing the dramatic first effects of climate change. The erosion of their coastline, the salination of their water supply, changes to patterns affecting agricultural production. More intense and frequent Pacific storms were devastating their lives. The impact on their society had already started to show. Young people were moving away, seeing no future in the island way of life. They were seeking work elsewhere. Their tiny island was literally dying.

She had just come from the summit, where she had participated in one of the round table discussions. During that discussion she had been asked to explain the impact of climate change on small island states. In the discussion that followed, another country (she didn't say which, but clearly a rich, northern country) had asked her in a matter-of-fact way: 'What are you going to do when your country is no longer habitable? Where will you go?' This question filled her with rage. The tone of it, the assumptions underpinning it, disgusted her.

As she spoke, everyone in the room was struck dumb. Many people were fighting back tears. She went on, her voice quivering with despair and anger: 'What was I meant to say to him? Will we ask New Zealand if we can annex a small part of the country and rename it?' The whole assumption that somehow her people, her island, and thousands like it, were dispensable, and that people involved in negotiations could talk in that way, shocked us all. These people, these islands with their beautiful coral reefs, had become somehow disposable. And for what?

This deep frustration and sense of betrayal by the whole human family is palpable when you talk to many of those on the front lines of climate change. On another occasion, I was asked to help at a major event organised by Mary Robinson, who has become a key leader and inspiration in the fight against climate change. She organised a big event in Dublin Castle to explore how Ireland could play a role in tackling hunger in the world and fight climate change. Instead of merely inviting the usual political leaders and policy makers to this important summit, she decided to reserve almost half the places for those in the front lines. She invited over a hundred people from vulnerable communities to come and share their testimony of how climate change was affecting them.

The helpers and the communities spent two days together preparing for the event, getting to know each other and sharing our stories. Sitting down with these people and taking the time to listen and learn about what climate change is doing to them was a real eye-opener. I sat at a big round table next to a representative from the Inuit community in the Arctic Circle. Her story was heartbreaking; her community was being devastated by the thawing ice sheets. As a result, multinational companies were now moving in to take advantage of this changing world they had helped to create. Climate change had made it viable for them to now prospect for minerals and oil under the pristine surface of the Arctic, due to the thawing permafrost. She was campaigning to protect her community lands and well-being from the potentially disastrous effects of the mining industry. Yet it felt like a losing battle.

I met with a community leader from Barbados. With his colourful Rastafarian hat, big smile and droll, chilled out Caribbean accent, it felt like we should be ordering a rum and coke as we chatted. Yet his story was as far from the image of tropical paradise as you could imagine. He told me about the effect that climate change and biodiversity loss were having on the fishing communities on the islands. The traditional fish they caught have literally disappeared. Piers and harbours were being destroyed by more frequent storms and violent hurricanes. Whole communities risk losing their livelihoods or homes. At some point, they would have no option but to leave. But where would they go?

For me, these people became emblematic of all those activist Naomi Klein says are 'in the sacrifice zone'. These are the people, mainly men and women who have little opportunity to be heard in powerful circles, who are bearing an almost unbearable burden. They are often already facing vulnerability from other injustices, they have done the least to cause the problem and they face the brunt of the climate crisis. They usually depend directly on the land and the weather for their livelihoods and can no longer do so.

In my work with Trócaire, time after time communities all over the world have spoken to us about climate change and its destructive effects: 'The weather is changing'; 'We plant and the rains don't come, then they come too hard and destroy our crops'; 'We have had to sell our cows to survive'; 'Our families cannot survive on this land anymore'; 'We will have to move'; 'Where will we go?' The crisis has become very real. It is not something they need scientists or politicians to tell them is real – they can feel it in their bones, they can see it around them. When you spend time with them it becomes all too real.

Trócaire commissioned its first report into the impact of climate change on the people it works with in 2007. It carried out a survey of all its partner organisations and asked them simply whether changes in the climate were having an impact on their work. The survey showed that 95 per cent of the partners reported this to be the case. This survey was followed up in 2010 with more rigorous scientific research.

Together with the Institute of Development Studies in the University of East Anglia, Trócaire then carried out a two year study, following families in Honduras, Bolivia, Kenya and Malawi to understand whether climate change was having an impact on them.[1] The results were very clear. Families that are already experiencing poverty are made more vulnerable by climatic shocks like droughts and floods. It limits their options. Some were already having to 'adapt' by leaving the land and migrating to cities. Some who were not able to move were facing increasing periods of hunger. Climate change was already making a difficult situation even worse.

These effects became all the more pronounced in *Feeling the Heat*, the report detailing research that Trócaire carried out with climate scientists in Maynooth University in 2014.[2] Given the evidence it had gathered on the human impacts of climate change, Trócaire decided to look more deeply into the impact that climate change was currently having, and predicted to have, on countries in which it works. The team in Trócaire worked closely with climate scientists to examine the human effect of climate change in terms of economy, health, housing and so on. The scientists worked to help unpack the technical language within the scientific report and translate the findings into arguments that decision-makers without a scientific background could understand. The findings were not surprising, but deeply worrying.

The report looked at five countries, all of which have very low levels of emissions and high levels of poverty. It showed very clearly that across the board, the first effects of climate change are already being felt in many areas.

In the Philippines, the report found that over 60 per cent of the population live by the coast and are highly vulnerable to climate change. The coral reefs, which are the lifeblood of fisheries in the country, will be badly affected, and sea levels are forecast to rise and some low-lying islands may be completely submerged. Over half the land area of the Philippines is at risk economically from hazards such as floods, typhoons and earthquakes, meaning around 76.6 million Filipinos are vulnerable to the economic effects of natural disasters. Of course natural disasters have always been a feature of life in the Philippines, but projections of climate change indicate that the Philippines is expected to experience a significant rise in temperature and increased rainfall variability, with the highest increases projected to occur in major agriculture regions. Climate change, therefore, presents a systemic challenge to the country's efforts to address poverty.

Kenya, another place where Trócaire has worked for decades and seen significant progress in poverty eradication, faces an equally precarious future. Among the major challenges it faces is the prospect of its second largest city and the region's largest sea port, Mombasa, being severely flooded. A sea level rise of only 0.3 metres would submerge an estimated 17 per cent of Mombasa. In the best-case scenario based on current projections, sea levels will likely rise by 0.26–0.55 metres by the end of the century but could be 0.52–0.98 metres if emissions are not reduced.

Moreover, in Kenya, as for many other Sub-Saharan African countries, climate change is already leading to health crises. Changes in temperature have meant that malaria-carrying mosquitoes can now reach large cities such as Nairobi, where this was not a major problem until now. Women and children are particularly vulnerable to malaria. There is a growing scientific consensus in Kenya that recent malaria epidemics in the western highlands are connected to changing climate conditions. In Malawi an increased incidence of floods and droughts has had an adverse effect on the health of citizens. Such extreme weather events are associated with higher rates of infant mortality due to malnutrition and chronic illness associated with malaria, cholera and diarrhoea. As far back as 2006, the Malawian government has been reporting that the incidence of malaria is expected to increase and spread to previously cooler zones as temperatures increase.

The purely economic costs of climate change on the poorest countries are enormous. While the costs of adapting to a new climate are only emerging, an initial estimate of immediate needs for addressing current climate-related issues, as well as preparing for future climate change for Kenya in 2012, was put at $500 million per year. The cost of adaptation by 2030 will likely be in the range of $1 to $2 billion per year. Ethiopia, which has suffered severe recurring food crises over the last decade, could lose in the region of $2 billion as a result of losses in agricultural output due to rainfall variability.

By 2050, it is estimated that climate change could reduce the size of the Ethiopian economy by 8–10 per cent per year and increase variability in agricultural production by a factor of two. Already, increasing temperatures mean that attacks of crop pests are reported at higher altitudes. For example, the most significant coffee pest, the coffee berry borer, had never been reported in plantations above 1,500 metres until fifteen years ago. Now it is affecting these valuable Arabica coffee crops. This will have implications for livelihoods based on coffee production, with small-scale farmers likely to be hardest hit.

We simply don't hear these stories enough. Unless someone works in the field of international development, or has a reason to really delve into the issue of climate change, the fact that this is happening right now could pass them by. In the West, perhaps we are so bound up in our urban lives, dominated by fleeting media, we can't really make sense of what is happening. It is difficult to join the dots. Increasingly bizarre weather events like fires in Fort McMurray in Alberta, Sonoma in California and central Portugal or the most powerful hurricanes in the Caribbean are still for the most part seen as that – bizarre, isolated events. Even ex-Hurricane Ophelia, which wreaked damage on Ireland in 2017 and the freak snows of the Beast from the East in early 2018 passed from the headlines remarkably quickly. We are struggling to make sense of the link between these events and draw any conclusions. In the meantime, as we struggle to make sense of what is happening, those on the front line of climate change continue to pay the price as their lives and livelihoods are sacrificed.

NOTES

1. Trócaire, *Shaping Strategies: Factors and Actors in Climate Change Adaptation*, 2012, www.trocaire.org/resources/policyandadvocacy/shaping-strategies-factors-and-actors-climate-change-adaptation
2. Trócaire, *Feeling the Heat*, 2015, https://www.trocaire.org/resources/policyandadvocacy/feeling-the-heat

HEARTBREAK _____

When you hit your forties you suddenly realise that thirty years is not such a long time. You start to look back and forward in equal measure. When I was at school I remember thinking that the Second World War, which we all studied, was ancient history. I put it in the same category as the Fall of the Roman Empire or the Great Fire of London. In reality, that war had only happened thirty years beforehand. I see this in my own children too – they ask about things that happened before they were born as if they took place in ancient times. Time seems to become increasingly elastic as you get older.

I was reminded of this at a children's workshop I ran. I had been invited to come to talk to a group of twelve year olds and answer their questions about climate change. The twenty or so children were part of a summer camp and I had been asked to explain climate change to them with a view to getting them involved in local action. It was a rare sunny summer's day, and I could tell no one wanted to be indoors. The classroom was stuffy and the sun teased us through the slatted blinds. We sat in a big circle and I was introduced.

As I was talking to them, I suddenly remembered how it felt to be their age. Back then I had a sense that my life was an open book. I felt that I had all the time in the world. The idea of decades passing was something too remote to even grasp. I believed that I would live forever. Despite many challenges in my own childhood, it seemed like opportunities were at my fingertips if I worked hard and took advantage of them.

I grew up in a hopeful generation which firmly believed in upward economic and social progress. It was based on an unshakable belief that things will continue to get better in the long run. Despite personal setbacks, in fact, every generation since the war has seen the most phenomenal increase in living standards and life expectancy. They lived through the rationing of the 1940s and 1950s, and various cycles of boom and bust which have happened ever since. Yet despite the ups and downs, Western societies have held on firmly to the belief that humans are on a relentless trajectory of progress. It is like there is an invisible line which transects time and leads only in one direction – upwards. So long as society holds on tight to the relentless march of time, and technological advance, the future is bright. The phenomenon of climate change really challenges that techno-optimism in a very deep way – as I found out quickly when I sat with that group of children.

After they listened to me talk a bit about climate change and all the facts and figures, there was silence. The awkwardness went on as everyone shuffled and looked at their feet. The group leader tried to coax the children to think of some questions, sensing their unease. Then one of them raised his hand and asked, 'Can you describe what the world will be like when we are grown up if the adults don't change what they are doing?'

The question was like a bombshell. As I sat there thinking how to answer I felt profoundly sad. A big part of me wanted to gloss over the question and to respond by saying something a little less harsh than the reality. After all, who am I to worry them?

I felt I had to speak the truth to them. Children growing up today have a right to know. So, I started to describe what the world would be like in the 2040s and 2050s based on the predictions of climate models. I started by explaining that there are many, many uncertainties in those models. They are based on such complex probabilities that it is hard to envisage the future with any degree of certainty. But the uncertainty is about *speed, sequencing* and *degrees* of change – not about whether changes will happen. It was a very hard conversation to have but they listened intently.

What I told them was this: 'If nothing changes, if we continue at current rates of emissions and we don't find any "magic bullet" technology which can safely capture all the greenhouse gases, if we refuse to change our economies and societies and don't slow down, then global temperatures will be well on the way to rising by 2°C, and on track to rise by up to 3°C by the end of the century.'

'What does that actually mean?'

'Well, scientists have said that the maximum amount of warming which is "safe" (and even that is not safe really for those living in vulnerable situations – safe perhaps for us in rich countries), is 2°C. We have already warmed up by over 1°C. Beyond that threshold things start to deteriorate rapidly.

'Beyond the 2°C threshold – which scientists now say we are 95 per cent likely to breech by the end of the century – we are in uncharted territory. We don't know which of the changes will happen first – but there are already very disturbing signs of warming oceans, melting ice caps, increasing interior temperatures, melting permafrost in sub-Arctic regions, which will have a domino effect on the world around us. Changes in the ocean currents, in particular, which have already started, will cause further changes in weather patterns.'

The children's faces were impassive. Either my stories were going over their heads or they were quietly trying to make sense of what I was saying. Or maybe they just wanted to be kids and allowed to go outside and play.

'The result of all this, while uncertain, will be widespread devastation to some parts of the world and disruption to others. For example, large swathes of the interior of Africa, which are already facing desertification, will become virtually uninhabitable. People who live there will have to find somewhere to survive. Those who can move, will certainly move. They will move to the big cities like Nairobi, Addis Ababa, and Khartoum, increasing the size of these cities and generating more tensions and sprawling slums. Many will move northwards and make the perilous journey to Europe and the Middle East, which will increasingly close its borders and seek to fortify itself as a safe enclave for its existing population.

'By the 2060s, when you are middle aged, dramatic sea level rises will already have taken effect. Predictions vary greatly, but based on continuing emissions rises, scientists put sea level rises between twenty-one and seventy-one centimetres, with a best guess of forty-four centimetres average. The main cause of this will be melting ice caps and expansion of oceans due to heating up. Many of the world's coastal cities, including Dublin, will be faced with some very difficult choices. Certain areas will be uninhabitable and by necessity will need to be abandoned to the sea. Like the end of the last ice age, new coastlines will be formed – but this time the oceans will have to absorb the detritus from our coastal cities and towns. Many millions of people across the world will be made homeless. The oceans could be effectively destroyed due to the resulting pollution as the remains of human settlements and industry are consumed by rising seas.

'Farming will be put under serious stress and could collapse under the force of the increasingly erratic climate. Many staple crops, which feed the bulk of the world's population, such as rice and maize, will be badly affected. Water shortages and sea level rises are just two of the major problems that farmers will face. The cost of food could increase dramatically as governments seek to stockpile supplies to protect their populations. The open trading system we have come to take for granted as globalisation, which enables us to get easy access to food all year round, will almost inevitably change.

'Mass migration, like that witnessed across Europe over the past decade, and new refugee crises, will be common place in the future. Climate refugees, seeking to escape rising seas, loss of land and livelihoods, will become common place. Yet it is unlikely that many countries will open their doors to these people. The result could be an increase in tensions, conflict and violence, as the few (the wealthy few) seek to protect themselves from refugees seeking new homes.

'Hurricane-type storms will become normal for your families. A warming atmosphere means more energy and moisture in the air. This means the potential for larger and more intense storms increases. What have been seen by us as once-in-a-lifetime events will be a very regular occurrence. Hurricanes and typhoons will be common place, even in Western Europe.'

I paused for a minute to see if they had any reaction to this bleak scenario – but still none came. Then one of them courageously asked a question.

'Do you think this will really affect us here in Ireland too? Surely the grown-ups will sort it out before it gets too bad. My dad said they are only a few years from developing a new nuclear technology – then it will all be sorted. We will have a free source of energy like the sun.'

I didn't want to be pitted against one of their dads! I told them that there are indeed some promising technologies, but bigger changes need to happen too. I realised I had said more than enough. So, I asked them how this made them feel. 'Angry,' one said. 'Scared', 'upset', 'powerless'. I could tell that the weight of my words had had an impact. Perhaps too much. I felt as if I had just burst their bubble. So, I stopped. I asked them to work in pairs and try to come up with some potential solutions. How could they try to address this problem?

But I could have said much more. I didn't have the heart to say how their hopes and dreams could be directly affected. When I think back over how I spent the last twenty years, I realise that one thing will differ greatly for them. They will most likely not have the same privilege of travelling that I have had. I have travelled all over the world, visiting every continent for work and pleasure. I have done so in relative safety. I have enjoyed the amazing experience of setting off with a backpack thrown over my shoulder with no idea where I was going to lay my head down that night. I have thumbed lifts across islands and slept under the stars in the Amazon. The sad reality is that this generation of young people might be the last one to enjoy such freedoms. In the future, if emissions continue to rise and the climate changes, these basic freedoms will simply not exist anymore.

I could have also told them that when they graduate, their job prospects could well be much more precarious if climate change continues unchecked. As far back as 2006, the serious economic implications of climate change were clear. Professor Nicholas Stern, a prominent British economist, published the most wide-ranging economic analysis of climate change.[1] The report said that the cost to the global economy of runaway climate change would be at least 5 per cent of GDP each year by mid-century. His predictions are now thought to be conservative. If more dramatic predictions prove correct, the cost could be more than 20 per cent of GDP. Sharan Burrow, Secretary General of the International Trade Union Confederation, said in the run up to the climate talks in 2015 that 'there are no jobs on a dead planet'.[2] Her words have become something of a mantra for those who highlight the links between economics and climate change.

The truth is that climate change is not good at all for the economy and especially not for workers. Open economies, where goods and people can move around freely, thrive where people have relative security. Investments tended to thrive where there is greatest political stability and security to make long-term choices. Climate change, by its very nature, is disruptive and unpredictable. It makes it far more difficult to predict how investments will perform. It will inevitably lead to many assets – physical assets in particular – being damaged or even destroyed.

In fact, climate change is now ranked as one of the most serious risks facing the global financial system.[3] The Governor of the Bank of England, Mark Carney, talks of three specific risks to the global financial system and specific investments. First there are actual 'physical risks' to property, machinery, trade, infrastructure, people or other physical assets. The increasing frequency and intensity of weather events due to climate change means that there will be greater insurance liabilities.

As the risk grows year on year, this will increase the cost of insurance premiums to protect these assets. As floods and storms damage property or disrupt trade, their value will be affected. Some homes and properties on coastlines, for example, will need to be abandoned. They can be protected once, twice – or maybe even up to ten times – but as the frequency of floods and storms increases, insuring these properties will become impossible. The cost of rebuilding will far exceed the value of the asset. The idea that this situation could be one facing entire coastal communities, let alone cities such as New York or Cork, makes you realise the scale of the economic problem the world will face. What happens to all the jobs? Who pays for the displacement?

Secondly, there are 'liability risks' for those who are involved in damaging the climate. Those affected by climate change could effectively seek compensation from those they hold responsible. What if those children today, who will face a very uncertain future due to climate inaction, were to sue their governments for failing to protect them? Such claims could come decades in the future. Given that governments and corporations have known about climate change and its causes for decades, and knowingly continued to pollute, they cannot claim ignorance.

Finally, Mark Carney talks about 'transition risks' in the economy. Transition risks denote the uncertainty around the cost and timing of bringing about a lower-carbon economy. Since the Paris Agreement in 2015 there is a growing belief that governments are attempting to tackle climate change. This is resulting in rapid changes in policy, moves towards different technologies. As the whole world simultaneously seeks to move from an economy which is highly dependent on fossil fuels towards one which has minimal impact on the climate, this could 'prompt a reassessment of the value of a large range of assets as costs and opportunities become apparent'. In other words, as the world switches to new, clean technologies – such as solar energy – old ones can suddenly become defunct and virtually valueless.

History is littered with examples of such transitions which render previous technologies, especially in the field of energy, redundant. When the motor car was introduced in the early twentieth century, horses and carts were at their peak; but once motor car technology became more sophisticated, it saw spectacular growth. Similarly, the compact disc replaced the cassette, the MP3 replaced the CD and streaming services replaced the MP3. Generations of technology leave a long trail, but the superior technology always wins in the end.

The major problem is that such transitions, on the scale predicted due to climate change, are rarely simple or easy to predict. They can happen dramatically and without warning. The proprietors of the old technology will always hold out as long as possible. They point to the many false starts as companies engaging in new technologies try to capitalise on the innovation. Energy transitions are littered with failed start-up companies. Many people will lose a lot of money. Some will get very rich. Like the Grand National, there are many starters, many fall on the way, and only a few make it to the finish line. The winner often remains frustratingly difficult to predict.

The period these young children will be growing up in will be marked by profound economic and financial upheaval. What has been taken for granted (in financial terms) has suddenly become uncertain due to the reality of climate change and the ongoing doubts over government response. If it is not taken seriously, then the costs of doing too little, too late will continue to mount. Lord Stern's worst-case scenario of economic collapse may well come to pass. In such an economy, there will also be growth sectors – those of 'disaster capitalism'. People who are responsible for picking up the pieces of the climate crisis will see some spectacular growth – security firms, military hardware, humanitarian response. But the demand for other jobs which depend on a sense of global stability and economic well-being will dwindle.

What I also didn't tell that group of children is that this economic shock and the political upheaval it brings, mean that they will grow up in a much more dangerous and unpredictable world. All the evidence points to climate change leading to fierce competition over increasingly scarce resources. In particular, access to fresh water in certain parts of the world will become a major flash point for conflicts. The long-term effects of climate change, such as recurring droughts and floods, which lead desperate people to flee for their lives will become a major catalyst for conflict. A major study in 2017 cited twelve 'epicentres' of new conflicts across the world that are already emerging due to climate change.[4] The study points to the inter-connections with pre-existing problems in the world. The authors of this report point to the ways in which water is already being used to control local populations, either through diverting scarce water sources, as Islamic State did when it seized the Mosul Dam in Iraq and threatened to demolish it. In the future, water shortages may lead to new inter-state wars. Egypt, for example, has already threatened Ethiopia with air strikes over damming the River Nile. Climate change is creating new problems too, bringing the world into uncharted territory. There are no international norms for dealing with the reality of small island states disappearing beneath the waves. For many, this is fast becoming a reality they are facing. For now, perhaps, the number of people affected remains relatively low. In the future, however, it could lead to climate refugees on a mass scale.

And then there is the cost to the beautiful planet itself. Climate change will affect the habitats of so many awe-inspiring, beautiful places and creatures, leading to a mass extinction not seen since the last ice age. Predicting exactly which species will become extinct and when is very difficult. Some studies have claimed that between a quarter and half of all species could be lost by the end of the century if climate change is not addressed. By the time these children are in their fifties, so many species could be gone from the wild. What is beyond doubt is that as the planet warms up, the stress on different species increases and more become endangered. The list of species at risk is long and depressing. Among those high on the endangered list are sea turtles, penguins, polar bears, seals, possums, butterflies such as the Sierra Nevada blue, and all types of coral. Researchers at the US National Oceanic and Atmospheric Administration have predicted that by 2050 almost all coral reefs around the world will be afflicted by bleaching-level thermal stress each year, with the resulting loss of many species that depend on this rich ecosystem. Humanity may not depend directly on all of these species for its survival, but their loss due to human actions is profound and real. It is so sad to think that the next generation may be the last to see those animals and plants in the wild. Their children will need to read about them in picture books. What a tragic loss.

My thoughts returned from these unspoken harsh truths, back to the children in the workshop. They had finished working in pairs. As always, they came up with a wonderful list of very sensible ways that climate change can be addressed demonstrating to me, once more, that this is not rocket science. Solutions exist. They painted a picture of a different, brighter future: Turn off the lights to conserve energy, put solar panels up, use less and recycle more, cycle and walk more, become a vegetarian, buy less stuff, love all the animals, buy an electric car. As we finished up, one little girl came towards to me. She told me she wanted to work in nature conservation when she grew up, but she had a question which she thought was a bit silly: 'Are grown-ups not doing anything to sort out this problem? Do they not care what will happen to us?'

NOTES

1. Nicholas Stern, *Stern Review: The Economics of Climate Change*, 2006, webarchive. nationalarchives.gov.uk/20100407172811/http://www.hm-treasury.gov.uk/stern_ review_report.htm
2. Interview with Sharan Burrow, Secretary General ITUC, *The Climate Group*, 17 December 2015, www.theclimategroup.org/what-we-do/news-and-blogs/there-are-no-jobs-on-a-dead-planet-sharan-burrow-international-trade-union-confederation
3. Mark Carney, 'Breaking the tragedy of the horizon – climate change and financial stability – speech by Mark Carney', *Bank of England*, 29 September 2015, www. bankofengland.co.uk/speech/2015/breaking-the-tragedy-of-the-horizon-climate-change-and-financial-stability
4. Caitlin Werrell, Francesco Femia, *Epicenters of Climate and Security: The New Geostrategic Landscape of the Anthropocene*, 9 June 2017, www.climateandsecurity. org/epicenters/

SLEEPWALKING _____

What is it that prevents us from facing up to this momentous problem? Why is it, when catastrophe is staring our children in the face, that so few people seem to care? That little girl had asked the million-dollar question. Many books and academic studies have been written on why people are ignoring climate change. Some, like George Marshall's excellent book *Don't Even Think About it: Why Our Brains are Wired to Ignore Climate Change*, examine the many reasons why we are so reluctant to face up to the issues facing our planet. Others, like *The War on Science* by Shawn Otto, take the viewpoint that there is a real and existing conspiracy on the part of those who hold power to stop us from actually taking climate change seriously. Naomi Klein's book *This Changes Everything* encapsulates both sides of this debate perfectly by examining the way in which climate change forces us to rethink our economies and societies. What all of these books and studies seem to confirm is that the reasons we are not acting with a due sense of urgency are very complicated.

Understanding this complexity can be challenging, but perhaps a good place to start is with a bit of introspection. What if we were able to get beyond the complexity and bring the issue right down to the personal and examine what is actually stopping each of us from taking action? It is something I have had to think about a lot. I think I have a pretty good idea what was stopping *me* before I grasped the seriousness of this issue. I have a pretty good idea about what is still stopping me from taking more radical steps to limit my own emissions and those of my family even now.

The truth is that most of us who live in relatively wealthy, temperate countries like Ireland simply don't see the problem. Not yet anyway. Today I got up, got the kids dressed for school, jumped in the car as it was raining and did the school run. I went into the office, picked the kids up, did a grocery shop, ferried the kids to various after-school activities, came home, made dinner, got the kids ready for bed, tidied the house, did the laundry, texted a few people, and checked Facebook. I then sat down and relaxed in front of the TV. At which point in this busy routine did I have a moment to even think about the impending climate catastrophe? The only thoughts remotely related to this subject involved choosing which coats the kids were going to wear in the morning or lamenting the fact that I had forgotten my umbrella. I probably exchanged a quick word about the weather as I greeted my colleagues. I may have noticed that it was raining more heavily than usual.

For most people, particularly working mums like me, this sense of relentless routine and constantly being on the go is perhaps one big reason why we don't even think about climate change, or any other major world problem for that matter. The lack of time and, perhaps more importantly, mental energy to entertain such massive problems and how they relate to our lives is a big stumbling block. These issues necessarily remain someone else's problem. Most of us like to think we are doing quite a good job at multitasking. At a minimum, we feel we are just about getting by. Each day has enough problems of its own! That 'getting by' takes a lot of effort already. It often means overcoming inner conflicts about our identity. How many times did I find myself going back to work in tears with an overwhelming sense of guilt about abandoning my children? I regularly decided to be a stay-at-home mum, only to realise that I couldn't afford to do it. The constant juggling and mental gymnastics that all of us face in one way or another is itself exhausting. At the end of the day it leaves me with a feeling of entitlement and a sense that I deserve one of the gold stickers on the kids' reward chart.

If this is the actual reality for many, if not most, parents today, then facing up to our own part in causing climate change is going to be a momentous task. Each of us is on a treadmill of work and family which is very hard if not impossible to step off. We choose the path of least disruption even when the science shows that this path makes us at least partly culpable for compromising the very future we want to create for our children. It is hard to acknowledge that by virtue of living and working in the societies we are part of, we hold a certain amount of responsibility for what is happening. At a minimum we elect the leaders, we give legitimacy to their decisions. We all continue to live our lives pumping out the gases which are causing climate change. It is a problem that leads to a Pandora's box which few want to open or a buried sense of guilt which nobody wants to face.

Finding ways to get the message about climate change and its effect on the next generation across to parents and grandparents in particular is hugely challenging. Given that so much of our lives these days is mediated by short marketing sound bites, talking about something as serious and complex as this crisis is challenging. In a world where uncertainty, doubt, claims and counter-claims abound, communicating truthfully about climate change – and having that truth accepted – has become nearly impossible. The problem is that making people feel uncomfortable or guilty about what they are doing has been proven not to work. Unless people take time to really understand the problem, it is hard to break through the sound bites and short attention span of media. Even the fifteen thousand scientists who issued a second warning to humanity on 1 December 2017 stating that the current climate trends are 'alarming' and that humanity is now on 'notice', hardly made the news.[1] Even if it did, the impact seemed inconsequential. The problem is that as the stakes get higher the voices of those who understand what is going on just become more imploring, shrill and at times accusatory. Desperation doesn't lend itself to 'softly, softly', subtle marketing techniques.

This communication problem effectively means that we are able to continue to live our lives as normal – even if that normality is increasingly interrupted by peculiar weather events. The vast majority of us aren't yet being forced to 'join the dots' and link what we see around us in terms of the weather patterns, the floods and the forest fires to climate change.

The difficulty in communicating climate change to a wide audience has led to much research. 'Climate communications' has become a new professional area and employs increasing numbers of people. Al Gore, environmentalist and former US vice president, set up a global organisation, the Climate Reality Project, with the simple objective of sharing the truth about climate change with the general public.[2] Others have formed programmes like 'climate ambassadors' to get the message out.[3] Some, like the Climate Generation Project set up by the great Arctic explorer, Will Steger, specialise in communicating the message to young people.[4] So far, these projects have trained thousands of people all over the world to communicate climate change to ordinary citizens. It is a mammoth effort to shift public understanding on climate change which must bear some positive impact. Yet it is still a drop in the ocean when one compares it to the vast marketing budgets of those driving our unsustainable consumption forward.

Even when that message does land, and individuals are educated about the science, it is still a long journey between understanding and action. A lot needs to happen to get that theoretical notion of 'this matters' to translate into action. The problem just seems too big for one person to make any difference. Few of us know where or how to start. Even with the best will in the world, deciding to just 'go green' overnight isn't an option for most of us. Many people simply don't have the means to make even the most basic sustainable changes. Most people can't just buy expensive electric cars even if they wanted to. We can't imagine up renewable electricity providers that don't exist. We can't afford to just retrofit insulation in our homes at our own expense. We can't magic up public transport where it doesn't exist.

The central problem is that so much of the change we need is dependent on others – on institutions, government, corporations. A way of life that damages our environment and imperils the future has become part and parcel of our social and economic system. Our societies and economies, particularly in the West, but increasingly around the world, are structured around increasing material consumption. The notion of ever-increasing private consumption – or 'created need' – underpins the very structure of our daily lives and our whole society. It is driven by a multi-billion euro marketing industry. Marketing companies have long understood that in the modern world, shopping is not simply something we do to fulfil a functional need. It is an activity from which we derive a meaning in itself. It is something many people do as a leisure activity. It fuels economic growth and we are told again and again that economic growth is the reason for our happiness.

And there would seem to be ample scientific evidence of a strong link between economic growth and happiness. Growing economies, we are told, are associated with increasing levels of comfort, satisfaction, better quality of life and higher life expectancy. Generally, trends show that as a country's economy is growing, people have more money in their pockets, they can pay their bills, service debts and do more of the activities that satisfy them. People can invest in making and consuming more and more stuff which in turn keeps people in jobs. The financial crash of 2008, which led to a sharp contraction of the economy and a long recession, is a reminder of what happens when growth stops abruptly.

There is strong evidence to support the view that the higher per capita income of a population, the higher up that country appears on the Human Development Index.[5]

However, the assumption that human happiness is *only* dependent on economic growth has come in for serious questioning. Professor Jeffrey Sachs, Director of the Earth Institute at Columbia University, has led a project to examine these links more closely. His work is based on the 'gross national happiness' measure of societal well-being which has been used since 1972 by the small landlocked kingdom of Bhutan in the Himalayas. The annual World Happiness Report has become an important counter-narrative to the idea that economic growth is the be-all and end-all when it comes to quality of life and human happiness.[6]

What this report shows is that human happiness is complex and at times contradictory. The relative importance of a growing economy matters differently depending on whether you are rich already or suffering from extreme poverty. For those who lack the basics in life, a growing economy is essential if it grows in a way that allows access to those basic necessities. If you earn a few dollars a day, then reaching $20 or $50 would make a massive difference to happiness levels. But as the level of income gets higher, the return on every extra dollar or euro in terms of happiness decreases. If you already earn $200 a day, earning $220 may not affect your overall happiness so much. In fact, this landmark report has shown that for wealthy societies, seeking to become even richer can lead to growing dissatisfaction with life and even rising unhappiness. When that happens against a backdrop of growing inequality and environmental destruction, both at home and abroad, the marginal gains in growth can be outweighed by their negative spill overs. As Jeffrey Sachs puts this dilemma: 'Poor people would swap with rich people in a heartbeat. Yet all is not well. The conditions of affluence have created their own set of traps.'[7]

Those traps become crystal clear when you try to break out of them. You suddenly realise that this damaging pattern of consumerism, of a use-once-and-dump mentality, extends right across our entire lives. Recently I was making dinner and I deliberately put all the waste created by that one meal to the side. A plastic meat container, plastic mushroom box, mesh for a packet of onions, box for pasta – and all that for just one meal for four people. Multiply that by the billions over weeks, months and years – you get my point. And our food packaging waste is the least of our problems. You could argue that a home-made lasagne wasn't exactly a 'created need'. We all need to eat. In fact, the lasagne did us for two nights so we didn't do too badly. Yet it made me realise how much packaging waste I was creating just by serving up a simple meal.

Another example many parents can relate to is what happens when household appliances suddenly fail. In our parents' day, the logical thing to do was to call the repair man. Nowadays, however, repairing goods is becoming much more difficult. You may want to repair the appliance but soon realise that doing so is nearly impossible. The costs of parts might be prohibitive or the parts may no longer be available. Local people with the technical skills to repair may not even exist. Or the technology has simply been superseded. This happened to me with an expensive digital camera. The camera is in perfect working order – but the digital memory needed to be upgraded. When I went to purchase a new memory card I was told that type no longer existed. It had been replaced by a new, more advanced format. Unfortunately that format didn't fit in my camera! My three-year-old camera had become antiquated.

These advances in technology coupled with certain manufacturing decisions – and the marketing industry – mean that even if I want to stick with my old washing machine, camera or TV for a bit longer it is really difficult, if not impossible. Economic globalisation means that markets have been flooded with cheap goods, making it more economical to produce and ship a kettle half way around the world than to pay a local tradesperson to fix a switch. Manufacturing companies, moreover, have long been accused of building in obsolescence into almost all electronic goods such as televisions, phones, fridges, washing machines, cameras, and so on. Whether this is true or not, the reality is that the constant, relentless march of progress compels us to buy something new when the old one is perfectly fine or requires a small repair. What we had becomes virtually useless so that we can purchase something new and feed the system. Giles Slade has studied the history of this phenomenon in great detail and concluded that obsolescence is central to the consumer-driven economy in the USA, and by extension the rest of the globalised world.[8] It is not simply a case of greedy corporations seeking more profits (though in some cases, like ink cartridges, it clearly is), but a structural problem in our economy. It has become a core industrial strategy to artificially limit the lifespan of goods to ensure that we keep buying. Companies need to make money so they incessantly feed the human desire for what is new, different, better. Like magpies, humans have a fascination for shiny new things. In many ways, they fool us into believing that what we have isn't worth repairing and is somehow inadequate.

This consumerist mentality is so ingrained in us that it is now mainstream culture. It inculcates every aspect of our lives and those of our children. None of us is immune to it, even those who are very conscious of it and try to resist it. For example, when I was pregnant and unsure about what I needed and what I didn't really need. There were thousands of bits of information to digest. I picked up pregnancy magazines and read books. I was anxious to do the very best for my new baby. In many ways, I was an advertiser's dream. I would listen to anyone who told me convincingly that I should buy their product, that it would help my baby and me (especially with sleep and feeding). I ended up purchasing a pile of stuff, enough to fill a playroom. I bizarrely ended up with four different contraptions to help babies sit at the table – a strap-on chair seat, a booster chair bag, a clip-on chair seat and a high chair. In the end I used one my mum donated which was sturdy and in perfectly good nick.

Most of this stuff was totally unnecessary. The gadgets of all sorts were used once or twice and then sat in a box. They started to clutter up the house. My sister, who had a baby shortly afterwards, couldn't believe the bonanza coming her way. As I loaded up the car full of nearly new baby gear for my sister, it occurred to me just how vulnerable to consumerism we all are.

Our children are easy prey for this way of thinking to the point that it is almost a form of propaganda. Before my youngest child could say his name, he could recognise the logos of major corporations such as McDonalds and Tesco. My boys don't watch much satellite TV because I got fed up with the relentless pestering from adverts. Instead they watch Netflix and movies, and they are allowed a little bit of time on YouTube Kids. Yet the marketing finds its way right into the heart of their lives. It is hard to escape. Entire programmes are now simply marketing tools for Disney and Pixar merchandising. You simply cannot get away from it.

Of course, not all of this is new. When I think back to my own childhood I remember watching TV and dreaming of owning a Sindy doll or bedspread. But what is happening now is on a whole new level. The scale and integration of the phenomenon, is so targeted, so inescapable and perfectly designed to turn children into little consumers. The children literally get hooked on the rush of adrenalin that comes with novelty – be that a Kinder Surprise, a Happy Meal or a birthday present. It has all the hallmarks of addiction. Countless studies, moreover, are warning parents of the addictive nature of the screens which mediate these new forms of marketing. They are as addictive as cocaine on a young child's brain.[9]

The industrial scale of advertising, which effectively becomes a way of life, means that entire families and communities become locked into a way of being and thinking which has consumption at its very core. Every mother and father knows what this means: mess and clutter. Never before have our children had so much, never before have they expected so much, and never before have they gotten so little pleasure from their toys. It is a cliché to say that children get more fun from the cardboard boxes in which the presents come than the toys themselves. In our house this is certainly true. The boys gain more pleasure from a free box and permission to make a mess in creating something than they do from plastic toys.

In so many ways, this addiction to stuff, increasingly mediated by technology, actually robs children of their own imaginations and ingenuity. As Sue Palmer puts it so well in *Toxic Childhood*, the clash between Western culture and our children's natural ability to think, learn and behave is resulting in a profound crisis. It is evident in the many statistics around childhood happiness, which show rising levels of unhappiness and mental health problems in wealthy Western countries in particular. At the root of the problem, according to Palmer, is a disconnection between culture and nature: 'In a nutshell our culture has evolved faster than our biology.'[10]

The craziness of the whole system becomes clear when our houses, our garages, our gardens simply become littered with unused toys and other gadgets. We feel compelled to buy more for others and give material gifts that, quite likely, the other person won't appreciate. We have to spend hours managing all this stuff, finding homes for it, selling it off or giving it away to charity.

Yet solving the climate crisis is not possible unless we deal with the issue of relentless consumerism. The problems are deeply interwoven. The scale of the task and the urgency of doing something to change all this can result in a feeling of helplessness. Changing things, however, is not impossible. In fact, my experience is that once I faced all of these issues seriously, I started to see the possibility of a different logic which many others had already put into practice. It is a battle which won't be won overnight and means delving deeper into what it is that makes us truly happy.

NOTES

1. William J. Ripple, Christopher Wolf, Thomas M. Newsome, Mauro Galetti, Mohammed Alamgir, Eileen Crist, Mahmoud I. Mahmoud, William F. Laurance, 15,364 scientist signatories from 184 countries, 'World Scientists' Warning to Humanity: A Second Notice', *BioScience*, 67 (12), 1 December 2017, 1026–8, www.academic.oup.com/bioscience/article/67/12/1026/4605229

2. *The Climate Reality Project*, www.climaterealityproject.org; *Climate Talk Ireland*, www.climatetalkireland.com

3. *Climate Ambassador*, www.climateambassador.ie

4. *Climate Generation*, www.climategen.org

5. Human Development Index is a UN measure of quality of life. It includes a measure of life expectancy, education and access to basic needs. 'Human Development Index (HDI)', www.hdr.undp.org/en/content/human-development-index-hdi

6. *World Happiness Report*, www.worldhappiness.report

7. Jeffrey Sachs, 'Introduction to the first World Happiness Report', 2012, www.issuu.com/earthinstitute/docs/world-happiness-report

8. Giles Slade, *Made to Break: Technology and Obsolescence in America*, Cambridge, MA: Harvard University Press, 2006.

9. Seth Ferranti, 'How Screen Addiction Is Damaging Kids' Brains', *Vice*, 6 August 2016, www.vice.com/en_us/article/5gqb5d/how-screen-addiction-is-ruining-the-brains-of-children

10. Sue Palmer, *Toxic Childhood: How the Modern World is Damaging our Children and What we Can Do about It*, London: Orion Books, 2015, p. 3.

EMBRACING THE EARTH ⎯⎯⎯⎯⎯⎯

The saving grace for me, amid the emotional turmoil of facing up to our children's and planet's future, has been finding a deeper connection with our beautiful earth and learning to love it like never before. It has become a deeply personal, spiritual journey. Since I started on this journey it is like I have fallen in love with the wonder of the earth we live on once again. It challenges me at the deepest level to wonder about our earth, our universe and the meaning of it all.

Most of my own childhood, in fact, was spent outdoors. Despite the difficulties our family faced growing up (or perhaps also because of them), we largely made our own fun. For the most part, at weekends and holidays, this involved leaving the house in the morning and returning when I needed to be fed or watered. Our fun usually consisted of building dens in the fields at the back of our estate from branches, stones and leaves. We invented all sorts of adventures which could go on for days or weeks on end. We were never bored.

My mum always nurtured a love of nature in us. Rainbows and sunsets were especially important to her – and still are. Whenever she saw a rainbow she would point it out and say 'it must be a sign'. A sign of what, you might ask? Of love, of hope, of peace I assume. The ancient biblical sign of covenant between the heavens and the earth that God gave to Noah after the great storm. I still remember one evening when I was five or six walking up a big hill with my mum, having visited my baby sister in hospital. There was a particularly beautiful rainbow, so beautiful it has stayed with me for decades. There were three luminous rainbows one inside the other. For Mum, those rainbows were a sign of hope which I didn't really grasp back then, but can imagine what they meant now. Her mother was dying and her newborn baby was sick in hospital.

Sunsets also meant a lot to us. Some evenings when I was growing up I would see the red glow reflected on the houses opposite and run outside to catch a glimpse of the red sky. Sometimes it seemed like the whole sky was on fire. At times, I would run out of the house and find a quiet space to watch the sunset. I would even quietly give a round of applause for the free show which nature had just performed.

As I grew up, I kept this deep sense of connection with nature, but over time perhaps that familiarity, borne from always being outdoors, waned a little. My life became one of sheltering from the elements more than facing them. As I grew to understand the momentous task facing our world and the fragility of this earth, however, I rediscovered a deep spiritual desire to reconnect with nature.

For me, one important step in that spiritual journey came while I was reading a letter on ecology written by Pope Francis in 2015, ahead of the Paris climate change summit.[1] Since he became pope, Francis has worked tirelessly to highlight the need to change our behaviour to tackle earth's ecological problems and especially climate change. He has done this through his own example and by his writings on very pertinent social and environmental issues. He has invited a new dialogue between people of faith and science, in light of what is happening to the earth. So when I heard he had written a letter 'On Care for our Common Home', I was keen to study it. As soon as it was released online I sat at my desk and read it carefully. His provocative letter, addressed to every person living on the planet, did not disappoint. The letter was rooted in the latest science about the climate crisis. It painted a dramatic and profoundly worrying picture of the situation our planet is facing, and our inability to grasp the magnitude of the task ahead.

Pope Francis starts by recalling his namesake, St Francis of Assisi, a twelfth-century Italian saint, who is perhaps most famous for his love of nature. I have always had a curiosity for that particular saint, also because he is my namesake (my middle name is Frances). Each morning, when I was a child, before we left for school, my mum used to pray the Prayer of St Francis with us: 'Make me an instrument of your peace ...'

In calling to mind St Francis' deep connection with the natural world, Pope Francis makes a very pertinent point: how we feel about the world around us – our emotions, psychology, attitudes – shapes how we interact with the world. Pope Francis calls on us to undergo an 'ecological conversion'. I was struck first by the title he chose for his letter, *Laudato Si'* ('Praise Be'). When it comes to tackling climate change, praising God is perhaps not the first thing that comes to mind. Like many climate activists, I am more inclined to curse the situation or feverishly organise everything and everyone for the battles ahead. Praising was fairly low down on the list of things to do before I read *Laudato Si'*.

Yet, the opening lines of his letter, when I read it for the first time, hit me between the eyes. The earth, our common home, is 'like a sister with whom we share our life and a beautiful mother who opens her arms to embrace us'. This image of the earth as our mother I believe could hold the power to heal the deep fracture we feel between ourselves and the beautiful earth we live on. The relationship which he describes is one that should be marked by warmth, beauty, love, intimacy, praise, joy, wonder. Instead, too often it has become a cold relationship marked by indifference, dominance, and even destruction on the part of humans.

At the heart of our problems today, Pope Francis seemed to be saying, is that we have forgotten the utter dependence we all have on the earth, our mother. In fact, Pope Francis says that Western society, in particular, has allowed itself to believe its own super human myth: that we are masters over the earth. As Pope Francis puts it, society has developed 'an irrational confidence in progress and human abilities'.[2] This is seen in our fascination with technological advances like space travel, our ability to travel right across the world, communicating simultaneously with billions of people, generating electricity from separating atoms, and even manipulating DNA. All of this is truly astonishing, but it has blinded us to a much more profound truth about existence: for all our scientific advancements, we remain totally dependent on one small planet. The health of that planet and its finite ecosystem ultimately dictates our own fate. In fact, for all we have mastered, our relationship to the earth is more akin to that of a child being fed by its mother. Without the fruits of our Mother Earth, the resources she provides, the living space she offers us, the elements we are made of, we are nothing at all. If we damage or destroy those life-support systems in our quest to master our own fate, then in the end we also lose.

Allowing ourselves the time and space to rediscover ourselves as part of that earth again, and to be deeply grateful for it, is our first and most important task. In fact, amid all the tasks of tackling climate change, perhaps this is actually at the heart of the challenge. We are far less likely to destroy something we love deeply, something we see as part of ourselves. If we really appreciate something, we do so not out of entitlement, but because we recognise it as a gift. We treasure it. We cherish it. No matter which God we might praise, the point is the attitude of gratitude and humility towards something bigger, something greater than us. And this isn't some romantic notion. It's not all about tree hugging. Saint Francis of Assisi is sometimes presented as a romantic character – a kind of medieval Doctor Dolittle talking to the animals and birds. Yet, as Pope Francis reminds us in *Laudato Si'*, he was far from romantic. This deep bond he felt with all of creation was something so real to him, so compelling, that it transformed every facet of his behaviour. It led him to reject everything that had to do with appropriating that nature for himself. What right did he have to 'own' part of this world that had been gifted to him? What right did he have to destroy it at will? He understood something which our mainstream culture, at least in the so-called 'developed'

West, has dismissed – that there is far greater happiness in
being than in *having*. Our interior being grows not so much
through affluence and accumulation, but rather by being free
from excess. There is such a thing as 'enough'. After our basic
needs are satisfied – after we have the security of a warm home,
nourishing food, healthcare, education – material wealth
does not bring contentment. It may bring some happiness
for a while, but not deep and lasting contentment. It is life's
other things that give meaning, particularly the experiences
we share with others in a bond of friendship and kinship, not
the accumulation of stuff.

Cultivating that deep gratitude for life, and especially
the natural world, is a precious gift in itself. Perhaps the
biggest mistake I make when it comes to trying to 'solve'
environmental problems is to think that somehow I have to
'do' something for the environment. I see 'it' as a problem
and myself as the fixer. As Pope Francis says: 'Rather than
a problem to be solved, the world is a joyful mystery to be
contemplated with gladness and praise.'[3] Of course there is
a lot to be done on many different levels to resolve the crisis
the world is in. Yet before jumping into problem-solving
mode, before mobilising one's energy to solve the problem
of climate change, there is a moment of reflection which
needs to take place. It is like a moment of conversion. That
'ecological conversion' for me meant something very simple
– recognising the earth as my mother and asking forgiveness.
Asking forgiveness for the times I have taken her for granted,
for the times I have forgotten to ask permission, neglected to
protect her, forgotten to notice her or appreciate her beauty.

This simple act, this change of heart, for me started to foster a much deeper appreciation for nature and all its many gifts. I started to observe the forces at work more keenly, and took notice of the changes happening all around me.

I was drawn to spending quiet time in nature every day, becoming reacquainted with the wonder of existence, like a child again. Amid the noise and busyness of life, it seemed that I was drawn into a never-ending process of discovery, into a beauty which was also divine. It was like my personal image of God became bigger, more expansive, more infinite – if that is possible.

Like my mum, moreover, I began to rediscover the mystery that we call God not as a magician sitting on a throne in the clouds, but as an infinite mystery whose existence can be perceived, unfolded, and experienced above all in what has been created. This God was in the flow of life and the patterns of death and resurrection we see in nature. I saw God in those strange coincidences which brought particular people, places and time together. From a kind of Christianity which had become a little formulaic and dualistic for me at times, my eyes started to open to nature as a window on the profoundest mystery which each person and every living thing shares – the ultimate 'Why?' And that 'why?' seemed to speak back in a million ways about the web of life, the 'hand' of Providence, of a benevolent presence which allows things to come into being and breathes life into them. That same Providence governs the cycles of birth, death and rebirth in an endless resurrection. The Christian mystery, in fact, shifted from being something miraculous, almost magical and at times irrational, to something which is imprinted in nature – from the smallest being to the greatest.

At one point near the start of this process I was invited to take part in a 'Take Stock' management retreat. From the outset, the retreat was quite different to the usual leadership training I had attended. Normally I associate management retreats with presentations, formulas and lists of things to think about. This retreat, however, could not have been more different. There were no timetables, no questionnaires, no icebreakers or awkward feedback sessions. The whole three days revolved around a simple question. Each of us was asked to imagine we had almost reached the end of our life (on our deathbed). We were to write about how we had spent the years that remain from that day onwards. It was a startling task which deeply affected all the participants in taking stock and re-evaluating our lives. It all revolved around Platonic idea: *the unexamined life is not worth living.*

A key part of the retreat involved going into the wilderness (which in fact was someone's back garden) and walking alone, deeply immersed in the present moment. I had to become as aware as I could of all the sights and sounds of the garden, using all my senses. I was to allow my attention to settle on one creature or plant. Observe it carefully, give it my full attention. My attention was drawn to a bumblebee which was hopping from flower to flower in a white jasmine bush. I could feel the sun on my back, the gentle breeze, the perfume of jasmine in the garden, the birds singing. The bee was working away. It was going about its business collecting nectar in complete harmony with nature. It wasn't paralysed by climate change or the threat of extinction facing its species. It was in the moment. The bee, it seemed, was saying something to me about my own life and my own concerns: don't fret so much that you can't live. You need to do your work, but do it serenely, without seeking to control the future. From that day I look at bees very differently. As a child I was terrified of them. I probably even killed a few in an attempt to enjoy a picnic. Now, I would even consider getting my own hive if I had some space.

Each day since then, I have taken a little more time to do this. To put down the books. To put away my phone. To go somewhere quiet. Into the wilds. To pick the most beautiful, sacred spot I can reach – whether that is the sunny corner of my back garden, a hill, a sea view, a park. Sometimes it simply means enjoying the view of a patch of sky from a café terrace. I have felt compelled to reacquaint myself with the earth in a childlike, wondrous way. I listen closely and hear the earth breathing, finding myself again in the beauty of the earth and appreciate all it offers each day.

For me, this process of reconnecting has been so powerful and life-giving. It has started to heal the rupture I feel inside me with our technology- and speed-driven culture. It has helped me to see that immersion in nature is essential for 'recovery' of ourselves and our planet. I can feel the healing power that scientists now tell us comes from trees. It has been life-changing. When I go into the forest I notice things. I give thanks for things – like the buds on the flowers and trees, the canopy dappled in sunlight, the uncurling ferns, and the squirrels in the trees. I don't take as much for granted anymore. I have such a deep desire now to learn the names of all the animals again and get to know more about who our cousins are out there in this beautiful shared planet.

I have found spending time alone in nature healing and restorative. Sharing this joy and welcoming others into this precious space has become even more special. This is particularly true when it comes to my own children. Like many, they have an amazing sense of intuition about nature. Once, when my oldest boy was five or six, we went down to County Galway to visit Granny. We took a walk along the winding bog road like we always did. It was summertime and the sun shone down, bathing the path in light. Inquisitive little insects were dancing around our heads. Bumblebees were drinking in the nectar of wild flowers. The flowers seemed like a rainbow growing up between the cracks of the cut away earth. I was struck by the intoxicating fragrance generated from what seemed like such tiny specs of dust. As we walked along, singing and dancing, we gathered flowers for Granny. Spontaneously, he stroked the grass, then looked up at me and exclaimed, his face beaming with love: 'Why Mummy, our Mother Earth has such beautiful hair.'

'Yes,' I replied, 'indeed she does.'

He bent down to pick another flower so gently. 'Do you think she can feel it when we pull it out?'

Nature is by far the best playground. This is something I have tried to nurture with my boys. Our favourite thing to do together as a family is to go for a walk in the woods. Near our house we are blessed with several beautiful woodlands. A walk through the forest is a big adventure for the boys.

118

They experience a sense of freedom and exhilaration which no human-made playground can imitate. Sometimes we go to a play centre if the weather is really bad, but there is simply no comparison between the way they play and relate to each other in the indoor plastic jungle and how they connect in nature. In the forest they run, jump and fall. They get cuts and scrapes, nettle rash and ripped trousers. They get covered in mud and come home soaking wet. Yet their faces are full of wonder and adventure. The green canopy towers above their heads. They wonder at the insects and the creatures all around. They can hear the creaking branches, unsure what is lurking round the corner. The sky can suddenly darken, the wind pick up and the gloomy forest fills them with foreboding. Nothing can compare to the wonder and atmosphere of being in the woods together!

On one occasion, we were walking through the woods. The light coming through the treetops seemed to turn the whole place into a magical green cavern. The boys ran off, delighted with the green space, climbing trees, picking up branches, wading through streams. I ran to keep up with them. We reached a resting point and one of the boys turned round to me and asked: 'Mummy, where did all this come from?'

'I don't know,' I answered. 'Where do you think?'

'Mmm ... I'm not sure.'

'Did you make it?' I joked.

'Nooooo.'

'Did Daddy make it?'

'Nooooo.'

'Did the Prime Minister make it?'

'Don't be silly!'

'Then who did make it? They must have been pretty amazing to make all this,' I said.

'Did God make it, Mummy?'

'Nobody really knows for sure. Maybe God made Mother Earth so we can all enjoy her beauty.'

'WOW! That's so cool. We better take good care of her then.'

The simple act of being outdoors and cultivating a deeper understanding of our world together has really enriched my family. It is amazing to think that such a profound change can happen at almost zero cost. I have realised that when we are out together we really start to notice things. We see all the beauty around us, but we also become acutely aware of the way in which humans are damaging that beauty. The children are the first to notice if someone has littered and they find it hard to understand how someone could do this. We often take a bag with us and pick up any litter we find on the way. In our own housing estate too, my boys have become known for their awareness of litter. One evening, a local representative from the Tidy Towns group even turned up at our door with two mini litter-pickers for the boys. They were excited and couldn't wait to get outside with their high-vis vests and pick up any litter around the streets. For them, this simple task had become a fun activity.

Spending more and more time in nature has another effect too: it has filled up our lives with such rich experiences, there is simply less time available to buy stuff. Since we are outdoors, all of us are less exposed to the marketing messages that feed our addiction to stuff. Of course, now and again we still enjoy going shopping and buying new things, but it has become far less important to us in the last few years.

Being outside in nature also gives me a sobering sense of my own fragility. It reminds me that existence is a fleeting state within a bigger cosmos. Often, we don't like to be faced with this reality – it may be why we sometimes shy away from deep experiences of nature. For me, there is nothing so awe-inspiring as looking up at the stars from a dark place. Several years ago I was visiting Argentina and I went with some friends to the Andes around Mendoza. One evening we decided to go out late at night to stargaze. We went high in the mountain pass where there was no trace of city lights. We lay down on our backs and stared up at the sky. At first it looked black – pitch black. There was no moon, no shadows. As we stared, however, the sky seemed to light up from within. From darkness, the sky suddenly seemed to be alive with stars. The star clouds of the Milky Way formed across the sky and thousands, millions, billions of tiny lights seemed to sparkle like diamonds. Far from being dead space – a blackness – the sky was alive with light and movement. Nothing seemed to be still. Shooting stars criss-crossed the sky. Satellites made their predictable paths across the horizon. And the whole sky seemed to be revolving as we tried to take in the scale and beauty of the heavens. As I closed my eyes, the darkness seemed to be speckled with spots of light. Opening them again, the stars appeared to be falling on us – the darkness had disappeared completely.

Nature presents us with such breathtaking experiences. Stargazing is just one example. Like standing on the top of a mountain or gazing out at the endless sea, it fills you with a sense of your utter insignificance in the grand scheme of things. We are but stardust. This infinite smallness, our fragility, is part of the magic of life. Lose that awareness – of the miracle of being alive – and you lose the perspective on everything. Recover that perspective, and everything else seems to flow.

With the publication of *Laudato Si'*, Pope Francis has become one of my all-time heroes on this journey. He has a rare ability to speak very directly about what is going on – to speak truth to power – but to do so from a place of love. It is a message which is uncompromising, but at the same time full of hope for a better future if we learn again to love and appreciate our earth for what it is – a home we share with all God's creatures. I was, therefore, thrilled to be invited to help mark the first anniversary of his letter on the environment at an event in the Vatican where scientists and activists considered whether any progress had been made since the letter was published. Pope Francis came to greet us and spoke of his deep concern because of how slow things are changing in terms of tackling the issue of climate change.

Then he cut to the chase: 'If we want to have an earth to pass on to our children, we need to act quickly.' Some have called him an alarmist. Those who understand the facts, including the vast majority of scientists, call him a realist.

NOTES

1.   Pope Francis, *Laudato Si'*: On Care for our Common Home, Vatican Press, 2015.
2.   *Laudato Si'*, 19.
3.   *Laudato Si'*, 13.

CHANGE AT HOME _____

Reconnecting with and deeply appreciating nature again was the starting point of understanding what I personally might be able to do to tackle the momentous problem of climate change. It has rooted me more in the physical reality of the earth in all its wondrous mystery. It has opened my eyes to how much we as a species take the beauty around us for granted. Ensuring that future generations, especially my own children, can enjoy the wonder of this planet and continue to benefit from her many fruits is what spurs me on every day. Yet obviously doing my part in tackling climate change is more than simply spending more time in nature. If only! That connection is as vital as the oxygen in my lungs. For me, it has become like a daily prayer. It may be the deep root of a change, but the change has to become action. Allowing that appreciation of nature to grow in me was the necessary first step, but the next – which continues to this day – is to address my own complicity in nature's destruction.

Finding practical ways to break out of the consumerist traps, untangle this mess and to start to form new habits, is essential. When it comes down to it, a substantial amount of the big numbers that make up CO_2 emissions I talked about in chapter five related to energy, transport and food can be traced right back to individual choices which each of us make on a daily basis. The vast majority of these are made at a household or family level. That got me thinking about whether there are key changes each of us can make while we 'hold out' for the bigger changes. Is there a kind of 'carbon diet' we can all go on to try and decrease our household emissions?

I discovered quickly that there are many tools online which can help in understanding where to start. One which I like a lot is called the Ecological Footprint Calculator.[1] This is a handy little tool which works out the impact of your travel, home and shopping habits on the environment. Its scope is broader than greenhouse gas emissions, but given the interconnections between different environmental issues and the wider economy, that's no bad thing.

I filled in the online questionnaire, which took fifteen minutes to complete, and it came up with a figure showing how many planets we would need if everyone on the planet was to have my lifestyle. It turns out that if everyone on earth lived like my household we would need 3.3 earths to sustain life. My date for 'using up' my share of resources each year is the 23 April.

I have to say when my number popped up, I was quite shocked and felt quite discouraged, even a bit embarrassed. Here I am, I thought, trying to raise awareness on climate change and my own ecological footprint is pretty significant. Until then, I thought I was already doing quite well. Our family lives in a very well insulated, A-rated home with solar energy. We recycle as much as we can and try to be conscious about packaging and waste. We try to limit unnecessary purchases, particularly clothes and gadgets. I walk to work and usually walk the children to school too. My husband uses public transport most days. We are certainly not perfect, but surely all this effort goes some distance to helping the cause I feel so passionately about.

In an attempt to understand what was causing us to have such a high footprint – and hence where I could make further changes – I started to play about with the tool online. I started imagining different scenarios, putting in different factors like limiting trips abroad, opting for a more fuel-efficient car and house, and so on. I wanted to really understand if there was something else I should be doing to bring our own impact on the environment down? Was there actually a way to get to living within our planetary limits?

Interestingly, besides having children, the calculator showed me that the single biggest problem was our house. It startled me to discover that the mere fact that I live in a medium-sized, concrete, detached house and use electricity, I am breaching my planetary limits. This wasn't such a good start! If we moved to a smaller semi-detached house with similar energy efficiency we could halve our household footprint. Of course we *could* change our house, but it would be a significant investment – and not feasible if you think everyone would need to do the same to have an overall impact on climate emissions. The need to build different homes would immediately offset any gains – and of course someone would also move into our house. It had already shown how complicated this issue is.

The second big issue turned out to be what our family was eating. I realised that if I changed the settings to vegan I could go some way to reducing our environmental impact. This seemed like a more feasible option. I cooked meat of some sort most days. As a family we seemed wedded to the notion that if there wasn't meat in it then it wasn't a proper meal. When I altered the settings to become a vegetarian, I realised that I could reduce our household footprint significantly. This certainly gave me lots of food for thought – but made me realise the uphill struggle I am going to face.

The third variable I played around with was foreign trips. I make three or four trips each year for work, one for a foreign holiday and one to visit family abroad. My husband has to travel most months for work too. It turns out that the number of annual flights (which until then I considered modest) was having a massive impact on our household's footprint score. If we halved the amount of flying hours, our score dropped accordingly. This was extremely challenging news to take on board.

Having calculated my household footprint, I decided to start making some changes – starting with the simplest ones within my control. Food choices was the obvious place to start. As I saw it, we had a number of choices: cut down gradually on meat consumption, become vegetarian or go vegan. As a mum with two growing boys and my husband to think about, the complexity of making these choices immediately became apparent. There was more than myself to take into account – but was it my place to make choices for everyone? Life is busy enough without having to make different meals for different members of the family! I anticipated a backlash.

Yet, as the main grocery shopper and main cook in the house, I felt that I had a big responsibility – and an opportunity – to make changes. On the food front, I took a gradual approach. I simply made the change to using less meat and the others have hardly noticed. The first thing I did was to choose one day a week which is totally meat free. In many ways, this reinvents an old Christian tradition of fasting and not eating meat on Fridays. Many other faiths still hold such feasts regularly throughout the year. Meat Free Mondays is a campaign which was launched by Paul McCartney and his family in 2009 to help people cut down on meat consumption. I found their website very helpful.[2] Going meat free one day a week has established the idea that it is possible to make changes. I then introduced another day – Fishy Fridays, where we try to only eat fish from sustainable sources. From there, I started to look at meat substitutes – halving meat in some dishes such as stews and replacing it with beans or another alternative. I quickly found it is possible to have a really tasty meal that everyone will enjoy with no or next to no meat! Over time I have built up my meat-free recipes and we have moved to two or three meat-free days, and no doubt others will follow. The main thing is that we have broken the link between 'a good meal' and eating meat.

We have also started to take a keen interest in where our food comes from. We have discovered wonderful local organic food producers and started to support them. We have also started to grow our own vegetables and fruit in the garden. And I'm convinced this is only the start. Making more sustainable food choices proved to be easier than I thought. But transport, especially flying, has proved more complicated. To be living within our ecological footprint, like everyone else, our family would essentially have to stop flying. Unfortunately this is just not practical if we want to continue doing our current jobs and living in our home. It is a fact of modern society – one of the 'traps' Jeffrey Sachs speaks about. Flying is part and parcel of work for both myself and my husband. I actually hate flying and find it pretty scary, but like many people today, I have based a lot of my life around my ability to fly. I have family in another country and without cheap flights, I would rarely be able to see them.

However, some changes can be made. As a family, we have made conscious decisions to go on holidays at home more often, partly due to environmental considerations. Discovering the amazing country we live in has been a revelation. When visiting family in the UK, we take the boat as far as possible. Yet as the children get older and want to explore the world, there is a deep desire to bring them to visit other countries and see the beauty of the world. Like many others, I plant native trees to make up to some extent for my flights. But I know deep down that such acts are really about soothing my conscience rather than saving the planet. It is clear that I haven't resolved this particular issue – but I believe that there is a serious need for an honest conversation around it.

While it can seem like an uphill struggle at times, some things can be readily done to live in more ecological households. As well as changing our food habits and cutting down on flying, we can all start to shift towards a more circular, local economy. This is where I have tried to make most changes at home to date and had most success. The circular economy is a simple idea. It involves thinking about our interactions with the economy as a circle, or cycle, rather than a straight line. When we buy something, we tend to think only about that 'thing' whilst we have it. We don't tend think enough about the before (production) or the after (disposal).

The circle is encapsulated in the idea of the three Rs: reduce, reuse, recycle. This is nothing new. It is now widespread and taught in schools. Nowadays, a few more Rs have even been added, such as 'revamp' and 'reclaim'. The latest R is to 'refuse' – refuse unnecessary goods, refuse packaging and so on. This idea of the Rs is really useful as it helps us to focus on just how much we produce, how we consume it and how we dispose of it. It stops us from thinking just about the short period in which we have a certain thing (for example a plastic bottle) and more about the whole lifecycle of the product. The major campaign on plastic waste in the oceans which took off following the Blue Planet II series demonstrates how urgent it is to shift our thinking in this direction.

That cycle is a more accurate description of our relationship with the things we create. Just because we stop using them doesn't mean they stop existing. The most pressing case of this relates to single use plastic items like bags, bottles, and straws. Most of this waste ends up in the oceans, turning them into what, in *Laudato Si'*, Pope Francis calls 'an immense pile of filth'. Reducing the waste in the first place is perhaps a bit harder, but still possible. When we start to talk about reducing waste we are beginning to touch on personal choices in the shops about what we purchase. Buying things that have

131

less packaging, for example, is always a good starting point. Avoiding any plastic containers is another way we can all help. There are excellent guides online and apps which can help guide purchases in more eco-sustainable ways.[3]

Thinking in terms of a circular economy has challenged me and the whole family to think carefully before we buy. Rather than making purchases on autopilot, we have held off buying something, looked to see if it can be bought second-hand, repaired or if we can make do. It has also led me to connect with like-minded people in the local community who are engaged in campaigns to end waste and encourage recycling. It has led to discussions about new local community initiatives such as repair cafes and revamping centres. Suddenly, from feeling helpless, many opportunities to build something new started to bubble up.

These changes are all important, but the footprint calculator also makes it very clear that many of the solutions which are urgently needed are beyond the control and budgets of most households. I can't just go out and change my type of house very easily to lower my footprint, I can't easily change where I work – most people commute to work not out of 'lifestyle choices' but out of necessity. When it comes to other items you might think we have more discretion over, for example how many big electrical goods we purchased, the same issue also appears. If my washing machine stops working shortly after the warranty is up and I am told it is a lot cheaper to replace the whole thing than get it repaired, or that the parts no longer exist as the model has been discontinued, then I know what I will do.

Other changes which require substantial investments can be planned for and happen over time if resources become available. For example, buying an electric car is now an option – but it is still a very expensive one. It is certainly a choice we will make when the time comes to get a new car. Installing renewable energy in the home, such as geothermal energy or more solar panels is also an option. Again, it requires a substantial investment of time, money and energy. Some government supports in Ireland are available to do this through grants for sustainable energy.[4]

While some people with large disposable incomes could certainly afford to be investing in these eco-friendly options, for the majority of people, even availing of the grants requires a substantial upfront investment that would be beyond their capabilities. If you need to balance getting food on the table, paying bills, renting accommodation, buying uniforms and books for kids going back to school, healthcare – the long-term ecological footprint of our households is far down the list of concerns.

Of course the risk is that we think that if our efforts won't solve everything then they are not worth doing. It is the issue that has dogged climate action at every level for decades. Yet here, I found Pope Francis very inspiring. He gives these efforts we make a name: 'noble acts'. They may not change everything immediately, but they have an immense value: 'There is a nobility in the duty to care for creation through little daily actions.'[5] He goes on to list many of the actions outlined above, saying that each of these small efforts can have meaning, acknowledging the sacrifice involved. 'We must not think that these efforts are not going to change the world. They benefit society, often unbeknown to us, for they call forth a goodness which, albeit unseen, inevitably tends to spread.'[6] He also says that such acts aren't peripheral to faith, but 'essential to a life of virtue' and 'not an optional or secondary aspect'.[7] In spiritual terms, the shift needed is best encapsulated in the word 'conversion' – a radical change of direction which is manifest in the every day. The effort to change, though hard and complicated at times, has a far deeper meaning. Above all, this effort has to be seen as part of a long re-education process. It is a process of becoming more acutely aware of the need to shift our whole mentality away from a materialist one based on consuming more and more, towards one which values the age-old idea that 'less is more'. Embracing sobriety, while embracing simplicity and frugality, are all important parts of the shift that needs to happen.

Using the Ecological Footprint Calculator was a good exercise for me. It showed me what steps I can take immediately to lower my own emissions and those of the family. Those relatively small changes, such as choosing to fly less, changing our eating habits and walking more, gained a greater importance for me and for the family. They also led us to think more about other consumer habits – such as buying less, simplifying what we use and reusing much more. It really started to shift our mindset and helped us to break out of those consumer traps.

Moreover, trying to make these changes at home confirmed for me the glaring gap that exists between individual action and what needs to happen on the bigger scale if we are to tackle this problem on time. It didn't lead me to despair – far from it – but to see my own changes as a starting point to seek out others who want to change things. It led me out into my own community and to ask 'who's in?' And that is where the exciting change started to happen.

NOTES

1.   'What is your Ecological Footprint?', www.footprintcalculator.org
2.   *Meat Free Monday*, www.meatfreemondays.com
3.   For a list of the top six apps see www.ecowatch.com/6-mobile-apps-for-sustainable-and-ethical-food-shopping-1882012802.html
4.   For example, www.seai.ie/grants/home-grants/better-energy-homes/
5.   *Laudato Si'*, 211.
6.   *Laudato Si'*, 212.
7.   *Laudato Si'*, 217.

STORY-CHANGERS _____

Each small change I make to live more sustainably is a living reminder of my connectedness – of our need to accept limits, appreciate nature, live lightly. Yet, however valiant my own efforts, the cold hard truth is that no amount of personal change will actually save our children from climate change at this point. If that was to have happened, by now our individual household changes would have grown into new ways of living shared by all. Certain ways of doing things would have become totally unacceptable, and even illegal. But given the short time frames remaining to get our emissions down and the powerful forces working against this, small changes, if done in isolation, are simply not going to get us there on time. Something more is needed to bring about a step change.

This conclusion is troubling, demoralising and deeply depressing. I would truly love to write something different that is more motivational and uplifting. But I also know that there is a real need to accept this reality in order to use the short window that remains to make the difference our children need. While some change is possible on an individual level, many of the major causes of global warming are beyond our direct control. That does not mean that our individual actions, like all the changes I am trying to make at home, are futile – far from it – but they are not enough.

Whether we like it or not, the choices we make in our lives are mediated by institutions at local, national and international levels, be that by governments, businesses or international organisations. Those institutions have the power to set the path and control the really big decisions which have an impact for generations to come. It is those decisions that will determine the success or failure of tackling climate change. If decisions are being made and policies set today to drill for oil in the Arctic, invest in new coal mines in the USA or exploit vast carbon sinks like tropical forests, then my own efforts to stop eating meat or travel less pale into insignificance.

Unfortunately, those who have a lot to lose, as I have already said, such as the fossil fuel industry, have shown that they are prepared to put up a significant fight to preserve the status quo – no matter what is at stake, even their own grandchildren's future. They are prepared to throw everything at this fight, and have seemingly endless amounts of money to do it. Indeed, it seems that there is no depths to which some won't sink in order to ensure the prevailing 'business as usual' approach remains. The power of vested interests in delaying action on climate change, and denying that the problem even exists, is really significant. The denial of climate change has become a very serious, multi-million dollar enterprise. It preys on postmodern societies, mediated by social media channels, where opinions and facts become hopelessly interchangeable and are treated with equal merit.

Tackling this could seem like a David and Goliath struggle for ordinary parents, grandparents and citizens who care deeply about the future but have limited resources. In the face of all this, individual actions to reduce emissions can only take us so far. The big question today is whether there are ways that those individual actions can be leveraged to bring about a truly massive change really quickly. Is there a shortcut we can use to bring about a massive cultural shift? Given how this post-truth, damaged political system works right now, and the time lag that exists in decisions being made, is there any way to move things quickly?

The good news is that in our highly connected world, there are many ways in which we can shift things quickly if we so choose. The important thing, however, is to recognise that bringing about the big change we need involves more than isolated individual actions. It involves re-imagining entire communities, societies, economic systems. Essentially, it involves starting to make changes at home and also reaching out from our families, and connecting with people near and far. In this chapter, I will look at what this means at a local and national level – and then move on to how that change is building into a truly global, planetary movement.

Above all, starting to really address climate change requires us all to do one thing the Irish in particular are quite good at: *we all need to become storytellers*. The biggest obstacle we have today is that the stories we keep telling ourselves about progress, the future, wealth and economic growth no longer work. We have to start telling different stories based on the truth (the climate science) and using this to spark our imaginations about what a better society, in harmony with nature, looks like. We need to share the little stories of positive action far and wide (and this is where our individual change actions become important). Starting to build a different story has a tremendous power, as I have seen from the many climate conversations I have been involved in. Since I started my own journey on this topic I have had hundreds of conversations with people. Conversations have a ripple effect – they change the story we are telling ourselves and empower us to begin a new, more realistic one.

Conversations, moreover, based on these stories of change can take place everywhere: in homes, places of worship, schools, sports clubs, workplaces, universities, pubs, societies – everywhere – since the causes and impacts of climate change touch everyone and everything. Every type of community can talk about it and every person has something they can contribute.

There is a simple format for such conversations that works well too. It all starts with building a shared understanding of the science – climate change is real – and how it happens. The conversation then moves on to the human activities causing it, and finally on to what can be done and how to get there. Such conversations can be painful – since the impact of climate change affects some more directly than others. There can be anger and tears as people absorb this new understanding of the world and the future. They start to place themselves and their loved ones in that future. For some, there can be a deep sense of grief and betrayal. Above all, however, these conversations give rise to a renewed sense of common humanity and purpose – a commitment to work together.

Over the past few years, since I really embraced the need for change, I feel like I have been swept up in a moving tide of people who want to talk and change the story. I have been up and down the country and beyond talking about climate change and the need to act together. I have given talks to the most varied groups: mothers' groups, parish groups, interfaith leaders, business communities, academics, older people, nuns, priests, schoolchildren and financial experts – the list goes on and on! It is really interesting to see how different groups of people are beginning to realise that climate change is something that matters to them too. While they may not see themselves as environmental activists, they see themselves as people who share a concern for our common home – and that of their children.

Such conversations generally do centre around the facts about nature and on a deep concern for those facing climate change now, for our children and grandchildren. Such conversations always result in action because they draw us closer to what really matters. They help to reconnect us with what truly gives us meaning and makes us happy. They can transform those little actions, for sure, but also generate a new sense of being engaged, active citizens who are able to make informed choices.

One of the most refreshing things about these conversations is that they often bring together young and old. Given the urgency of change, many older people today, especially active retired people, understand that it is not enough to appeal to youthful idealism to fight climate change. Appealing to young people is really important but unfortunately, this generation

141

will not have the luxury of time. They have few resources right now and by the time they reach positions of authority it will be too late.

Older generations have a critical role to play in generating conversations and changing the story. In fact, given the time frames involved in achieving the change needed, everyone is needed. Everyone can participate fully, albeit in distinct ways. Resolving climate change now means acting together across generations to protect our children, and young people and their children. It is a unique and in some ways beautiful opportunity to rebuild a sense of intergenerational solidarity around a common issue. Tackling climate change provides a way to build up the human family in ways which embrace people from all around the world, but also those who have no voice and are yet to be born.

Sharing a different story of our place in the world can become the springboard for inventing the kinds of communities, economies and societies that can help us to be resilient to climate change in the future. It opens up entire communities to see how things could be different – to use their imaginations in creating sustainable ways of life which are good, perhaps even better, than what they currently know. They discover that lower consumption does not mean low quality of life. On the contrary, it is entirely possible to live well and sustainably if we act together.

Moreover, there are many examples of sustainability already in existence that can offer insights into creative solutions. Many prototypes, such as the Global Ecovillage Network,[1] have been in existence for the past thirty years or so. Like most prophetic voices, they were perhaps ridiculed by some at first as utopian, new age experiences, but are now widely accepted as important experiments in sustainable living. They are far from perfect, but they provide working prototypes of communities that are prepared to take a commitment to living in harmony with the environment seriously.

Many such local groups are emerging right across the world. In Ireland, as in other countries, communities have started to come together to think about how they can work to lower their collective carbon emissions. They very quickly see that doing this together – rather than just struggling on their own – is extremely empowering. They reconnect their communities on things that matter and feel a deep sense of purpose in realising that there are thousands, even millions of people across the world thinking and doing the same. They are able to connect with those people and together start to build a global community of people who are making the change – as well as demanding it. They are generating the solutions with their creativity.

Some of the wonderful groups leading the way on this are 'transition towns' – a global network set up to spearhead the action needed to make the transition to a low carbon society. These community groups come together in a given locality and share their story. They then start to examine how to build more sustainable local communities and how to lower carbon emissions. They have recognised that given the systemic nature of the problems we face, such as unbridled consumerism, it is essential to work together over time to educate people and come up with creative, practical solutions. The situation can't change overnight but requires a transition period involving individuals, businesses, community groups and local government. This movement has been hugely successful and groups are emerging across Ireland and beyond. One of the most established groups is Transition Kerry which runs many public events and links in with a number of community initiatives in Tralee, Kenmare and Dingle.[2] They are working on issues of food, biodiversity, energy, transport and community resilience to climate change. Like many rural communities, particularly in remote coastal areas, they are increasingly conscious of the growing impact of extreme weather and the need to adapt.[3]

When communities come together and start to think about sustainable futures, other things can happen too. They start to realise the significant assets and potential within their community. Suddenly from thinking they had few resources, they discover ways to lower carbon emissions and build stronger, more resilient communities. In the UK, and many other places, there is a strong movement to foster genuine 'sharing economies', harnessing the amazing power of social media as the enabling technology. Some of the most exciting new initiatives are in this area. The sharing, collaborative economy is a potential antidote to consumerism with many practical applications, especially at a local level. It is based around the simple idea that, by and large, there is enough 'stuff' in the world already, particularly in wealthy Western societies but we aren't using it efficiently enough. We can share it freely. In fact, the growing phenomenon of 'solidarity networks' was the topic of my PhD on the 'Economy of Communion' back in 2000.[4]

We have reached a point where there is no need to continue to extract more from the earth or keep manufacturing more and more things. Our economy needs to be repurposed and reoriented towards appreciating the value of things we have already, sharing them more, and circulating them when we are finished with them. If we really think about it, many of the goods we 'need' might not be needed just for us. Our possessions can be exchanged or donated. They can become 'ours' – or common property of a group or community. Why is it, for example, that every house in a street needs to have a lawnmower? Why does everyone need to have a tent? Why do we all need to buy such a vast array of baby and children's equipment which we use for such a short period of time? My quality of life won't suffer if the community lawnmower is stored in someone else's shed and I need to book in advance to use it. On the contrary, perhaps the necessary interaction with my neighbours might just enhance my sense of belonging and tackle isolation. They might even offer to cut my grass for me!

Yet thousands of other non-commercialised examples of this kind of sharing economy exist. They exist by and large where community ties are strong and resilient. This sharing, collaborative approach to economy also extends to the creation of shared spaces to grow local food and even generate local energy. Local community lending and bartering initiatives are great examples. The Library of Things is a project which has taken off in the USA and Canada, but is also appearing in Europe. Public libraries become the repository not only for books and information of all sorts – but also for useful objects we all need once in a while. In a sense, it is a form of rental, except guaranteed by membership in the library. It has breathed new life into many libraries, which were floundering in the shift from printed to digital books. Once again, the library has become a central source of shared knowledge and now, goods.[5]

The idea of a sharing economy, moreover, fits neatly with the idea of a 'circular economy'. If goods are shared more, then production and consumption levels of new goods are reduced. People reduce, reuse, revamp, repurpose, and where necessary recycle what they have. The possibilities are endless. FoodCloud, for instance, is a not-for-profit Irish company that helps to redistribute perfectly good food nearing its sell-by date from supermarkets and wholesalers to those who need it.[6] It feeds people for free and also prevents food going to landfill and reduces emissions. The 'freecycling' movement combines sharing and recycling.[7] It has become a powerful way to share or give away goods people no longer need, while providing much needed goods for those who have little income. Examples include numerous repair cafés and workshops across the world.[8] At these cafés, people can learn long-forgotten skills of repairing goods or bring them for others to repair. They do so in a way that brings communities together and tackles isolation and loneliness.

Online and physical community spaces have been emerging over the past decade to enable this kind of transition to happen. These initiatives, and many others, provide real solutions to our problems of consumerism and carbon emissions. They have been bubbling under the surface for several years. Now, they are ripe for a major leap forward. Social media, despite all its flaws, has enabled like-minded people to form new communities locally and then connect globally. It has led to local community climate action blossoming all over the world.

This is great cause for hope. Starting these conversations at a community level is the key. It is those changes at a community level that start to shift the political dial. All politics is local, as they say. There is absolutely no doubt that conversations result in a shift in attitudes, behaviours, and start to shift policy decisions. Only an educated, engaged citizenry that understands how our choices will affect the next generation will be able to give those in power the permission to make the changes we need. There is a perception in the political world, certainly in Ireland, that doing something about climate change will not win any votes. According to politicians I have spoken to myself, climate change is not on people's minds when they go to vote. All the choices involved seem difficult and deeply bound up in local politics – whether it is building more wind farms or stopping traditional practices like peat fires. It is only when people start to recognise that resistance to change is putting the future of our children on the line that perhaps we will manage to adapt our behaviour. Some countries, like Scotland, have undergone major shifts in public attitudes as a result of extensive public conversations on climate change. It has made it possible for those governments to introduce measures that would be considered by some as fairly radical. Big changes can happen quickly when whole communities of all sorts come together and focus on a shared challenge.

Ireland's record on translating commitments on climate change into real reductions in greenhouse gas emissions is not great. It faces a very particular problem in that a significant amount of its emissions do not come directly from fossil fuels but from the agricultural sector. Ireland has a vested interest in the beef and dairy industry and this has dominated the conversation about tackling climate change for the past decade. It has cast a long shadow over every other important conversation that should be happening in relation to energy, buildings, transport and so on. It has often become a battle between the rural and the urban populations. In the past year, however, it has embarked on a number of innovative experiments involving participatory ecological contract building.

The Citizens' Assembly was set up in 2016 to consider a number of key issues facing Irish society. One hundred citizens were selected at random to participate in a very engaging process which involves deepening their understanding of an issue and then proposing changes to existing laws and policies which are needed to address it. Climate change was selected as one issue which should be addressed. Over the course of two weekends, the citizens were offered expert inputs on climate change and Ireland's response. The inputs were based on the science and free from bias from campaigning or interest groups.

The outcome of the sessions was truly remarkable. On the last day of the assembly, the citizens almost unanimously voted to implement some very significant changes to policy. These included scaling up supports for retrofitting houses, implementing renewable energy schemes (including allowing people with solar panels to feed back into the grid), more supports for public transport and cycling, and perhaps most significantly, putting in place a proper 'carbon tax'. This tax, they proposed, would include a cost which truly reflects the impact of the beef and dairy sector on the climate.

The experiment has its limitations. Its outcomes are not binding and will now be considered by Ireland's parliament, the Oireachtas. However, the unanimous agreement on the need for urgent action is quite striking – and in stark contrast to the often confused messages which circulate in the media. This experiment shows the importance of educating citizens in the climate debate. When citizens were provided with the information they needed in a reflective space, over four days, with few distractions, they came to their own conclusions. None of the citizens were climate scientists, but like myself when I did my school project thirty years ago, they could grasp the facts readily. Climate action became something essential and urgent.

As well as bringing these proposals directly to the Oireachtas for consideration, the assembly will now be followed by an even more arduous task. Over the next two years, people up and down the country are going to engage in similar processes – at regional and local level – to try to understand climate change, how it relates to them, and what needs to be done. The ambitious National Dialogue on Climate Action will seek to mobilise the whole population in changing our understanding of what climate change means. Importantly, it will involve conversations between different generations on their shared future.

NOTES

1. *Global Ecovillage Network*, www.ecovillage.org; *GEN Europe Global Ecovillage Network*, www.gen-europe.org/home/index.htm
2. *Transition Kerry*, www.transitionkerry.org
3. *Transition Network*, www.transitionnetwork.org
4. Lorna Gold, *Sharing Economy – Solidarity Networks Transforming Globalisation*, Farnham: Ashgate, 2004.
5. 'Library of Things', *Sacramento Public Library*, www.saclibrary.org/Services/Library-of-Things
6. *FoodCloud*, www.food.cloud
7. *FreeCycle*, www.freecycle.org
8. *Repair Café*, www.repaircafe.org/en

PLANETARY MOVEMENT_____

Having conversations and becoming more educated about the climate reality we are all facing is absolutely key. It opens up the space in public debate for honesty and it can spark many imaginative solutions at a local level. Yet over the past decade as I have delved deeper into these issues, and especially in conversations with politicians and climate scientists, it has become all too clear that even this is not enough. In the face of many vested interests, dealing with climate change also involves action on a political level. Laws need to change. It means building an organised campaigning movement on a scale that the world has never seen. It has to become a planetary people's movement.

The genesis of a planetary people's movement has been all but unnoticed for several years. It has been emerging in the shape of many campaigning groups on different issues – global poverty, disarmament, human rights groups, tax justice campaigns, as well as environmental groups. All these diverse groups have their own specific interests, but in the space of a few years have started to come to the realisation that without a safe climate, everything else they aspire to is put at risk. Moreover, there is a growing recognition that the same political and economic forces preventing action on climate change are also leading to other injustices and inequalities. The many battles for justice are interconnected.

To succeed in tackling climate change, these diverse groups of campaigners will have to grow into a truly global movement – and do so faster than any other movement in history. It needs to become a 'movement of movements' in which all people who care about the future take part. It has to reach across all divisions – whether nationality, faith, gender, interest groups, political opinions – in a singular effort to tackle climate emissions and enable a just transition. It also has to be highly strategic and adept at dealing with political threats that will inevitably arise.

The power of this new emerging movement in large part lies in the diversity of the people who make it. It resides in the sheer number of people it can mobilise to demand change. In September 2014, I was present in New York when an estimated three hundred and eleven thousand people joined together to call for climate action. I will remember the day forever. It was filled with colour, hope, joy and anger too. There was dancing, drums and music. People from all faiths and none were present – praying and chanting in whatever way they liked. Many families were present with their home-made banners sporting wonderful witty slogans. There were many indigenous communities in their traditional costumes – many of them at the front of the march to represent the first victims of climate change. At exactly midday, this boisterous posse fell silent. For a whole minute, we each bowed our heads and remembered all those who are voiceless in this epic struggle: those who have died because of storms and droughts; those who are yet unborn; the countless species who are now extinct. Then, like a tidal wave, the silence was broken and an almighty cheer rose from the back of the crowd to the front. In that single moment, it felt like the tide of history was shifting. The power was with the people that day and filled us with a sense of hope.

Such peaceful mobilisation is now essential to moving the mountains in our minds that stop us tackling climate change. In the face of the awfulness of politicians who refuse to take the crisis seriously, getting onto the streets and shouting is liberating. It is cathartic. It builds a new sense of shared citizenship, especially when these moments are happening right across the world. Speaking out at a political level – and becoming a political force – is more essential than ever in the face of resistance to change. New political parties and leaders need to rise up from this planetary movement with clear manifestos for a safe future for our children.

Faith groups are at the forefront of this growing planetary movement, and growing more vocal in their demands for radical change. Within the Catholic Church, one of the most exciting developments in recent years has been the emergence of the Global Catholic Climate Movement.[1] This movement started off in 2015 as a response to Pope Francis' letter *Laudato Si'*. It was set up by a young Argentinian man, Tomás Insua, who wanted the plea for climate justice in *Laudato Si'* not to go unanswered. Within three years, the movement has grown into a real force within the Church, involving a rapidly growing network of over seven hundred institutions worldwide and implementing education and advocacy programmes across the world. At the same time, it is working ever more closely with people of all faiths and secular groups to ensure that political leaders hear and respond to the voices of the poor and younger generations.

Many parents are also starting to take action on climate change for the sake of their children. In Australia, from nothing, in just two years almost eight hundred thousand women have come together to focus on tackling climate action for their children.[2] Their campaign involves individual commitments to live more sustainably, as well as hundreds of practical community-driven campaigns bringing individuals and communities together to highlight climate change. They are making waves globally and starting to engage many more people in climate campaigning. Why couldn't that million become a billion? Why couldn't it also include a billion men?

One big question is what the focus of such an emerging planetary movement should be if it wants to have a significant impact? The reality of climate action is that lots of different things need to be done at local and national levels – and they all have a degree of urgency. However, several truly global campaigns have started to coalesce into this planetary movement. They offer real potential and hope to bring about a major global shift. They are campaigns that every person can get involved in right now.

The first campaign is the 'fossil fuel divestment' campaign which I have become directly involved in over the past two years. This campaign focuses on the critical role that finance and investments play in causing climate damage. The campaign was initially sparked by two very inspiring men who I have quoted already in this book: Bill McKibben and

Mark Campanale. Coming from two very different worlds – environmentalism and international finance – they started to follow the money, to ask where the oil and gas companies were getting the cash necessary to continue to invest and to make profits. They realised that despite their own policies, many governments still held substantial investments in fossil fuel companies. Many were subsidising their activities. They asked why it is that banks and ratings agencies are still treating these companies as 'A-rated' when it is increasingly clear that if climate change is to be avoided, then these assets are worthless. They simply won't be allowed to burn their reserves. Working from the financial analysis and the emissions figures, they started to cast doubt on the long-term financial stability of investing pensions in oil and gas.

In the course of a few short years this line of questioning has grown and grown. It became a smoking gun. Climate scientists and financial experts like Mark Carney, the Bank of England Governor, started to recognise that tackling climate change is a financial imperative – hence those companies whose value depends heavily on damaging the climate could become a liability, and quicker than everyone thought. If multiple governments start to move towards embracing clean energy, in line with their climate targets, and the cost of producing energy through such means takes off, what will become of all that oil and gas on the books? What of the pension funds that are heavily invested in fossil fuels? There is a serious possibility that such investments, if not wound down in an orderly way, could become worthless, or near enough. Crashes happen when we least expect it, and what

McKibben and Carney highlighted is a carbon asset bubble like no other.

It is a movement unlike any other environmental movement that has existed before. It not only brings together those who are passionate about caring for the earth, but also financial analysts and fund managers, who accept the science for what it is, and understand its profound implications for financial investments. The only movement it can be likened to is the divestment movement which sparked the end of apartheid in South Africa. Back then, in the 1980s, huge pressure was put on all kinds of businesses not to do business with the repressive regime in South Africa. They cut their business and investment ties with the country helping to isolate it as a pariah state until full democratic rights were given to the black population. In Ireland, Dunnes Stores workers were at the forefront of the struggle and went on strike to prevent Dunnes from importing goods from South Africa. The international divestment movement against apartheid may not have single-handedly caused the end of the regime, but it certainly helped speed it up. Nothing talks more loudly than money.

Only a few years ago, the idea that investors and banks would be taking their investments out of Shell, ExxonMobil and other massive oil companies was unthinkable. After all, we all rely on their product to run our economies. Oil is so deeply part of our energy system, our manufacturing and our transportation that it is hard to see how we can be filling the tank with one hand and holding a placard to call for the end of fossil fuels in the other. Yet that is exactly what has happened.

As of August 2017, the global divestment movement has seen over seven hundred organisations from more than seventy-six countries with over $5.5 trillion of assets under management commit to stop investing in fossil fuel companies. Of this, $1.5 trillion has been publicly committed to invest in climate solutions. A quarter of them are faith-based organisations, but the list also includes prominent universities, banks and cultural institutions that have significant endowments.[3] Several governments, such as the Norwegian government, have divested their sovereign wealth funds from the most polluting fossil fuels. As this book goes to print, the Irish parliament is in the final stages of deciding on a bill which will see it fully divest its own sovereign wealth fund from fossil fuels following a successful public campaign. If this decision is taken, as expected, Ireland will become the first country in the world to make its sovereign wealth fund fully fossil free. What is certain is that as cleaner, cheaper and, hence, superior sources of energy, such as solar and wind, rapidly come on stream, this tide of divestment will only grow and grow. The tipping point for an energy transition is already here and the change will happen rapidly.

Everyone who has a bank account can get involved in the divestment movement. It starts with simply asking where your own money is invested – challenging your own bank to come clean on its connection to fossil fuel investments. Depending on where you live, fossil free banking options exist in ethical banks, but also in mainstream banks too. Credit Unions also offer an alternative to mainstream banking. Beyond that, you can start asking where your school, university, faith community, or local council have their money. Many institutions now recognise the moral and the financial problem posed by investing in this sector. If they agree to go fossil free, they can sign a pledge.[4] Each institution that publicly commits to removing money from the fossil fuel industry weakens their social license and their grip on power.

The other area where the climate movement is growing into a planetary movement is through bringing legal challenges. As predicted by Mark Carney, many law suits have taken place in the past five years, and many more are now under way. A common refrain among them all is the role of vested interests, misinformation and political capture of governments resulting in their failure to protect the rights of particularly groups, especially younger generations.

The first prominent case of this type came to light in 2015 when a group of nine hundred Dutch citizens held their government accountable for taking insufficient action to keep them safe from dangerous climate change. The 'Urgenda' case was won on 24 June 2015, the District Court of The Hague ruled that the Dutch government is required to reduce its emissions by at least 25 per cent by the end of 2020 (compared to 1990 levels) meaning that the Dutch government is now, effective immediately, forced to take more effective action on climate change. However, in September 2015, despite calls from scientists, religious leaders, business and the co-plaintiffs, the Dutch government decided to appeal the historic verdict. Urgenda is currently preparing its response and the hearing at The Hague Court of Appeal is scheduled for 28 May 2018.[5]

Despite this setback, the success of the Urgenda case has emboldened citizens and concerned legal professionals all over the world, giving rise to hundreds of similar cases. In September 2015, Sarah Thomson, a then twenty-four-year-old law student, took the New Zealand government's climate change minister to court over its emissions reduction targets. She stated these were unambitious and failed to reflect scientific consensus on climate change. The judge hearing the case agreed that inadequate action had been taken, but no sanction was necessary due to the fact there had been a general election in the meantime. The move forced the incoming government party to increase its 2050 emissions targets accordingly. Similar cases are now also underway in Belgium,[6] the USA,[7] Ireland[8] and across the EU.[9]

The Irish case, which is being brought by the Friends of the Irish Environment, won a significant victory in the High Court in December 2017 when the judge ruled that Irish citizens have 'a right to an environment that is consistent with the human dignity and well-being of citizens at large is an essential condition for the fulfilment of all human rights'. This ruling sent shockwaves through government, which took it as a signal that future legal cases could be pending. The same group is now taking the Irish government to court on behalf of a group of eight hundred Irish citizens over its failure to act in accordance with the Paris Agreement and its own climate law. The 'Climate Case Ireland' campaign was launched in May 2018 as this book was going to press.[10]

The impact of these legal cases is already being felt. Though change since the Paris Agreement has been slow and faltering, over one hundred and ninety-five countries have taken some form of climate action since that agreement. Many have amended or instituted new laws to tackle emissions. In under three years, the legal landscape has shifted dramatically. The threat of future legal action has a cooling effect on questionable climate policies and denial in many parts of the world. Moreover, if the trend of successful cases continues, it could be very significant indeed. It would radically change the playing field and expose those who are robbing the next generation of their future.

When it comes to campaigning, recourse to legal action, however, is seen as a last resort for a simple reason: it costs a lot of money, expertise and time. It can be fraught with blind alleyways and dead ends. Almost always, whoever has the most resources to hire the best legal teams – and to keep the case going and going and going – has the greatest advantage. Despite this, many of the new generation of environmental lawyers working on such public litigation projects are young, dynamic and many are working pro bono.

The broader weight of this planetary movement has become critical to sustaining the divestment campaign and these legal suits. Their critical work is being largely 'crowdfunded' – using social media to enable ordinary people to donate.[11] Sometimes they become co-plaintiffs or simply give a little each month to ensure that this critical work is able to continue. Given the protracted technical costs involved, financially supporting this essential legal work is perhaps the single biggest way that almost every person can help change the laws which are preventing climate action. Imagine if every parent and grandparent gave even the cost of a cup of coffee to support a legal battle? The truth is that there are more parents in the world than there are super rich. Far more.

What is most striking is that this new planetary movement is not restricted to one category of people, one age group, or one political persuasion. I have discovered that there are many hundreds, thousands, perhaps millions of people who are waking up to the peril our world is in. Some of these people are directly affected by the bizarre, erratic weather that climate change is already creating. It is biting so hard for them.

What strikes me about the people I have met is that very few of them would have naturally considered themselves activists or environmentalists. The overwhelming number share one thing in common: they have joined the dots and believe in creating a safe future for themselves, their children, their grandchildren. They want to safeguard the other species with whom we share this planet. They understand what is happening and want to work to make it as good as it can be. Climate change just happens to be one hell of a problem that we need to face together.

The fact so many different age groups are involved in fighting for the future, as I've already said, also offers a poignant opportunity for intergenerational solidarity. The largest global movement for climate justice, 350.org, includes older people and younger people alike. When it comes to protest marches, however, they have a custom which came about because of the desire of the older generation, particularly those in their sixties, seventies, eighties and even nineties. The custom is that the older people put themselves at the front of the protest. This is a tactical decision – but has a huge symbolic value too. It is tactical in that if older people are arrested, the impact on them and their families is limited. They may get a criminal record, a night in prison, but it won't have an impact on their future. If a young person gets arrested, on the other hand, they may have a record for life. They may not get a job. Their lives could be seriously affected. The symbolism 350.org give to this, however, is even more important. They see this decision to place the older generation at the front as a way of symbolising the acceptance of their own part in the future the next generation faces. It is something which Bill McKibben talked passionately about when he participated in Trócaire's public conference on climate justice in 2015. His contribution is well worth watching.[12]

Many older people, moreover, are increasingly prepared to take peaceful direct action for the sake of their grandchildren, such as Phil Kingston from the UK-based group Grandparents for a Safe Earth.[13] Mr Kingston started the group with some friends in Bristol after considering the risk his grandchildren will face. Together with friends from his retirement club, they started to mourn the earth by carrying out fake funerals in the centre of Bristol. They had a coffin, a eulogy, and placed flowers at the entrance to Barclays bank. They decided on Barclays as they wanted to draw attention to the way in which the bank was funding the fossil fuels industry through its investments. They gathered signatures and petitioned the bank to change its investment policy on fossil fuels. During one of their funerals, Kingston managed to get into the bank and staged a sit-in protest in front of the customers. Bank officials, as you can imagine, were not pleased and called the police. The video of this rather gentle, elderly grandfather being unceremoniously removed from the bank went viral immediately. He was arrested, cautioned and released. Their protest, however, goes on.

Another older lady I met was Elizabeth Vezina from a group called the Raging Grannies.[14] I shared a platform with her at a conference organised by the active retirement group, the University of the Third Age. Their conference had the most provocative title: 'What do we tell our children when they start to ask us why?' The conference had a range of presentations, including one from myself, on climate action and the need for change. It is fair to say that Elizabeth stole the show. She was dressed in an oversize floral skirt, had a flowing shawl and a large floppy hat with lots of silk flowers on it. Before she had even started her presentation, she had lightened the mood and had everyone in stitches. Nobody quite expected it! She sang her presentation, telling us all that she is part of a movement of grandmothers and older ladies who campaign for climate justice. They use stereotypes about age to generate surprise, humour and publicity for their campaign. These ladies walk around dressed like any other, but at any point they can coalesce in a flash mob and become a 'gaggle' of raging grannies. They put on their silly hats and skirts and start to sing and dance together for change. They write and sing about the issues they care about, like climate change.

In 2016, group of raging grannies took their protest to Capitol Hill. It was a colourful sight to behold and again, went viral on social media. Many of them were arrested for singing, which was deemed a disturbance of the peace, but again, their attitude is one of Gandhian peaceful resistance.

The reality is that all generations need to work together to tackle the climate crisis now. Young people bring an idealism and passion – because the future we are fighting for is theirs. Older people, however, bring a whole host of different attributes and experience to this issue which youth cannot. Importantly, many can bring actual financial and technical resources. Many others can bring expertise of all sorts and influential contacts. Everyone has a part to play in generating this movement for a safer planet for the next generation.

Perhaps the time has come for parents and grandparents all over the world to rise up and get angry. Each day, week and month mothers and fathers everywhere make sacrifices for their kids. We do the most extraordinary things to ensure that they feel loved and protected. We agonise over whether they are getting what they need to flourish and grow. Yet in our quest to do our best, perhaps we have turned a blind eye to the biggest threat facing our children. We have inadvertently allowed it to get to a point where our inability to take action is jeopardising their future.

Perhaps the point is coming where we have to embrace the sense of acute anxiety when facing climate change and translate it into a focus on what really matters for their future. Everyone can take action and everyone can start at least one conversation. Everyone can make small changes and share those widely to change the story. But given the time we have and the pressures we face, shifting the curve of emissions downwards will take us all engaging our minds and hearts and joining the planetary movement for change.

NOTES

1. *Global Catholic Climate Movement*, catholicclimatemovement.global
2. *1 Million Women*, www.1millionwomen.com.au/
3. See www.divestinvest.org for details of organisations that have divested from climate damage and reinvested in climate solutions.
4. *Fossil Free*, www.gofossilfree.org
5. 'The Urgenda Climate Case against the Dutch Government', *Urgenda Climate Case*, www.urgenda.nl/en/themas/climate-case
6. *Klimaatzaak*, www.klimaatzaak.be/en
7. 'Juliana v. US – Climate Lawsuit', *Our Children's Trust*, www.ourchildrenstrust.org/us/federal-lawsuit
8. *Friends of the Irish Environment*, www.friendsoftheirishenvironment.org/climate-case
9. *GLAN: Global Legal Action Network*, www.glanlaw.org
10. *Climate Case Ireland*, www.climatecaseireland.ie
11. Platforms like www.crowdjustice.com are used to raise money for cases.
12. Trócaire, 'Climate Justice Conference 2015: Materials', www.trocaire.org/resources/policyandadvocacy/climate-justice-conference-2015-materials
13. *Grandparents for a Safe Earth*, www.network23.org/gfase
14. *Raging Grannies International*, www.raginggrannies.org

EPILOGUE ROAR _____

I started this book with the image of a lioness tending to her cub, and the instinct to protect her young. For me, that deepest instinct to protect my own young has become a driving force in raising my voice on climate change.

My own life has been shaped by a sense of urgency which resulted perhaps from my own childhood experiences of loss and fire. The impact and trauma of fire remains long after the acrid smell of burnt rubber and ash has passed. It remains beyond the repair and redecoration of a house. It remains in a sense of urgency that readily spills over into panic. If my house is on fire, I can guarantee I will scream, I will roar – I will do things I thought impossible to save my children, my home and my belongings.

In fact, I believe that the only power that is capable of dampening down the flames now licking at our common home is not one vested in politicians or even in inspiring religious leaders like Pope Francis. Certainly, leadership and policy change needs to happen there too. But those changes are not possible – in fact they are unrealistic and unthinkable – unless they are driven and surpassed by a different force. That force is one which is so visceral and written on the hearts of every person. It is the capacity to put the lives of our children first and to do what is necessary so that they can thrive.

The poet Drew Dellinger captures this sense of anxiety over what to do and the journey ahead in his poem 'Hieroglyphic Stairway':

It's 3:23 in the morning,
and I'm awake
because my great, great, grandchildren
won't let me sleep
my great great grandchildren
ask me in dreams
what did you do, while the planet was plundered?
what did you do, when the earth was unravelling?

surely you did something
when the seasons started failing?

as the mammals, reptiles, and birds were all dying?

did you fill the streets with protest
when democracy was stolen?

What did you do
once
you
knew?[1]

At the end of 2015, I was invited to give the closing address at the climate march held in Dublin in November. The morning of the march we woke up to a hurricane-force storm. We set out for the march – my boys armed with own fantastic signs. One said: 'Yoda Says – Be a Good Jedi: Save our Planet'. The other's was a bit more pessimistic: 'Don't turn our world into a Death Star.'

As we arrived in Dublin I was amazed to see just how many people had turned out. Over five thousand people were there. I knew many of them from all my talks in parish halls, but there were many others I didn't recognise. There was a carnival atmosphere, despite the wind and rain, as we wound our way up to Kildare Street, outside Ireland's parliament. As I stepped up on the stage to give the final speech the boys didn't want to let go of me. So I let them come up too. With them by my side and pulling on my legs, my words seemed to resonate more deeply.

As the crowd fell silent, this is what I said:

We are here today because we understand what is at stake.
We have accepted the science.
It tells us our world is in trouble;
We have made a choice that we must act.
Each of us has come on a journey,
And now we stand together.
Shoulder to shoulder,
Speaking with one voice:
We want decisive action.
We regret that too much time has passed,
Too many empty words have been spoken,

Too many excuses have been made.
We just don't buy it any more.
Too much is at stake.

We are here today because we care.
In our hearts we know
Things cannot continue the way they are going.
We cannot continue to pump out polluting gases
from our homes,
our cars,
our factories,
our farms,
our lives
And act like it doesn't matter.
It does.

We are here today because
We believe another world is possible,
We are prepared to put ourselves on the line to build it.
We know that solutions exist
And we are demanding that our leaders step up.

We are here today, above all, because
We know that there are many victims in the climate crisis,
Yet their voices are silent in the corridors of power.

If we stop and listen, we will hear their cries …

They are the cries of people
on the tiny atoll islands of the South Pacific,
the drought-ridden plains of the Horn of Africa,
the flood-hit plains of Bangladesh …
People forced to flee as a result of conflicts and disasters
made far worse by climate change.
They are the cries of future generations,
our children and grandchildren,
who will face a future of insecurity and conflict,
a burden we seem content to place on their shoulders.
They are the cries of thousands of other species
with whom we share this beautiful Mother Earth.

Their cries should be deafening,
yet their silence speaks volumes.
Who will be their voice?
Who will speak for them in Paris?

We are here today because we want to ensure that their
voices are heard
in the negotiating rooms in Paris.
That their rights are recognised and protected
and the ecological debt we owe them is paid.

We are here today because
we know that we can only change and build this future
if we connect with each other and make our presence
felt.
'To change everything, it takes everyone.'

Neighbour with neighbour,
community with community,
leader with leader,
nation with nation.
Bridging divisions, healing divides.
Finding common purpose.

Right across the world today,
tens of millions of people are coming to the same
conclusion.
Our voice of hope is far stronger
than the one that tells us we are doomed.
And let's be clear: this is just beginning.
Paris will at best, open the door.
But we will not stop,
because our hearts tell us
that this world is worth fighting for.

The lioness in me had roared.

NOTE

1. Drew Dellinger, 'Hieroglyphic Stairway', *Love Letter to the Milky Way*, Ashland,
    OR: White Cloud Press, 2011; drewdellinger.org/